The Theory of Prime Number Classification

The Theory
of Prime Number
Classification

Zwide Mbulawa

Copyright © 2010 by Zwide Mbulawa.

Library of Congress Control Number:		2010915527
ISBN:	Hardcover	978-1-4535-9893-1
	Softcover	978-1-4535-9892-4
	Ebook	978-1-4535-9894-8

All rights reserved. No part of this book may be reproduced or transmitted in any form or by any means, electronic or mechanical, including photocopying, recording, or by any information storage and retrieval system, without permission in writing from the copyright owner.

This book was printed in the United States of America.

Front Cover
Bottom Graphic (in greyscale):
The random distribution of prime numbers presented in a graphical form.
Top Graphic (in color):
The conceptual structure (model) of the algebraic sieve that predicts the exact location of all pseudoprimes. The algebraic sieve contains all the vital information about the primes, even the concept of the twin-prime conjecture.

To order additional copies of this book, contact:
Xlibris Corporation
1-888-795-4274
www.Xlibris.com
Orders@Xlibris.com

CONTENTS

Preface 9
Route to Discovery 11
Research Concepts 21
Research Framework 25

SECTION ONE

1 Introduction—Existing Classification 29
2 Basis of Theoretical Research 31
3 Prime Number Decomposition 34
       3.0.1 Nondecomposing Primes 38
       3.0.2 The Classification Rules 40
4 Classification of Primes 43
       4.1 Distribution Analysis 47
       4.2 Properties of the Classification System 50
5 Prime Waves and Event Signatures 57
6 Prime Reactions 62

SECTION TWO

1 Introduction 67
2 Defining the Prime Space P_s 68
3 The Default and Standard Prime Space 71
4 Group Behavior of Primes 75
5 The Prime Number Count Equation 79
6 The Mean of Primes 83
7 Average Gap in a Sample Subspace 85

Section Three

1 Introduction 91
2 The Prime Family 92
3 The Discrete Prime Space 96
4 Shadow Family 98
5 Some Applications 99
    5.1 Twin-Prime Conjecture in the Prime Space 99
    5.2 The Prime Count Space Gap 100
    5.3 The Prime Count Function $\pi(x, d)$ 101
    5.4 Group Frequency Distribution Patterns 104
    5.5 The Concept of Gap in the Gap 109
    5.5.1 Types of Prime Gap Analysis in P_f 111
    5.5.2 Comparing Gap Accelerations 121
    5.5.3 Probabilistic Gap Analysis 124

Section Four

1 Introduction 131
2 Gap Theory of Prime Number Classification 132
3 Structure of Gap Relationships for Classification 138
4 Fundamental Gap Behavior—Spectral Lines 144

Section Five

1 Introduction 151
2 The Random Function of Prime Numbers 154
3 The Generation of Prime Numbers 162
4 The Multi-value Function for Prime Numbers 168

Section Six

1 Introduction 177
2 The Concept of *G*-Numbers 178
3 Conceptual Structure of the Sieve 181

4 Meaning and Interpretation 188
5 Internal Dynamics of the Algebraic Sieve 194
    5.1 Pattern Structures... 198
6 The External Dynamics of the Algebraic Sieve................. 201
7 The Primality Test 210
    7.1 Why Pseudoprimes Are *G*-Numbers............................ 216
8 The Algebraic Prime Space 222
    8.1 The Integer Product Law.. 228
    8.2 Generating Gap Base Zero 230
9 Prime Number Curve... 237
10 Conclusion 1—General Prime Generation....................... 239
11 Conclusion 2—The Sieve Format.................................... 243
12 The Twin-Prime Conjecture 246
13 Epilogue... 251

Notes 255
Index.. 257

Key Words: Prime number classification, twin-prime conjecture, prime sieve, prime patterns, prime roots, prime number generation, prime numbers, expository mathematics.

Private Bag 00184, Gaborone. Botswana
Telephone 00267-71845286
e-mail: zwide@hotmail.com

Preface

One may wonder at the use of the word "theory" in the title of the book. The context of the word results from the use of four different classification techniques that have been developed because each technique has its merits. Formal methodology may require that only one technique should be applicable, especially in the area of mathematics since there should be one result to describe one concept.

It is my observation that it is possible to develop ideas completely independent from an existing framework. This research is not based on any existing program as such and does not rely on established theorems and conjectures to define its methodologies (this is a weakness if you want to publish in journals). In fact, there is no theory of classification of prime numbers as such, though some classifications systems have been proposed. Such systems have been limited in application going only as far as grouping prime numbers.

The meaning of what a prime number is has also deeply explored. The origins of mathematics have been the counting numbers, and prime numbers were premised on these.

The ideas, concepts, structures, and theories developed here have been tested for consistency and applicability through computer programming—this was the ultimate laboratory. Therefore, software is available at zwideprimes.com to validate, substantiate, and illustrate conclusions stated herein.

Zwide Mbulawa
September, 2010

Route to Discovery

Imagine I was given as assignment with the following tasks:

1. Develop four different ways of classifying prime numbers.
2. Define an algorithm that will generate prime numbers and pseudoprimes according to gap rather than natural sequence.
3. Investigate to define the source of random behavior in prime numbers.
4. Show that prime numbers are linearly related.
5. Show that there is an infinite number of functions that generate the prime number set with its pseudoprimes.
6. Construct an algebraic sieve that contains all the relevant information about prime numbers.
7. Derive a composite test.
8. Derive a primality test algorithm.

This would have been an impossible assignment for me, firstly because I never had much interest in prime numbers. Secondly, I am not an academic mathematician, I see myself as an entrepreneurial mathematician doing escapades in the expository domain. So how did the journey begin, you may ask?

It began as a result of two activities. I was writing a book called *The Science of Mathematical Thinking* (complete but as yet unpublished). In this book, I was exploring the concept of meaning, how meaning is developed in theory building in mathematics, and the relationship to mathematical objects. The book was a response to Crowe's tenth law, which states that "revolutions never occur *in* mathematics."

Crowe's argument was that no mathematical system has been overthrown and irrevocably discarded, and I took interest in this discussion because of another previous research on a concept called ditation mathematics. This was research into why quaternions are not commutative consequently deriving a triplet product and commutative quaternion and making a case against the *1-2-4-8* number model. These two works led me to study how mathematical methodology develops and how meaning is constructed, and I paid particular attention to the work of Rowan Hamilton to study his thinking pattern. Another law that influenced my current research was Crowe's law 7, which he explains by stating that "'gifts arrive in wrappings which must be torn asunder in order for the gift to be used or seen." The gift is the new mathematical concept, the wrapping is the philosophy or conceptual doctrine. By considering this law, I became an object of experimentation to myself. I wanted to research into what the wrappings were, how do I tear them asunder as I develop new ideas? I was interested to see how I develop new ideas from an intuitive level to a logical framework and the interplay between the object-level and meta-level of mathematical thought. I also wanted to see the dynamics of relationships in expository mathematics with regard to the creative and innovative process, though innovation is not a common word to use in regard to mathematical thinking. The research on primes provided an opportunity as it was a completely new field to me. As a result, I recorded almost all the gestations of thought, failure, and success in pursuit of mathematical development and conclusion.

Firstly, the journey began as I was trying to use my experience of the Barclays ATM card. When I was given this card, I never memorized the numbers; instead, I found it easier to identify the positions of the buttons that I must press to access my funds. At that time, my code was 1346. So keeping the buttons constant, I kept on changing the position of the numbers and consequently getting a different code each time (see diagram). As you may be aware, when the mind begins to see interesting patterns, it becomes explorative, so I began to add these numbers to find out if there was going to be a particular result. It was out of this exploration that my mind perceived a pattern in terms of having two numbers being prime coming together to define another number; thus, I began to conceptualize what I called the prime pair, and this resulted in the pursuit and construction of the prime root classification of prime numbers. The lesson here is that discovery can be inspired by any activity—it may not necessarily be traced to sequential phenomena.

THE THEORY OF PRIME NUMBER CLASSIFICATION

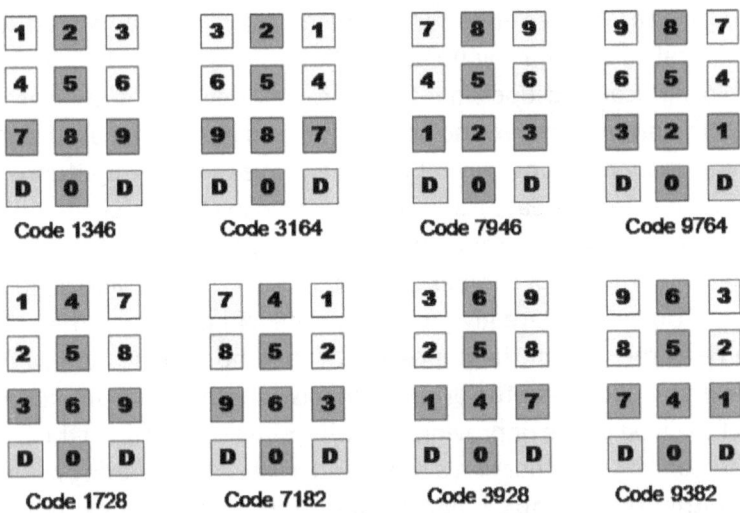

Fig-A. Code system that made me conceptualize the prime pair.

But as I have said, I had no interest in prime numbers; my focus was researching Crowe's assumptions and laws on revolutions in mathematics. And I was trying to find the connection between meaning and revolutionary thought processes. But this generated intense interest.

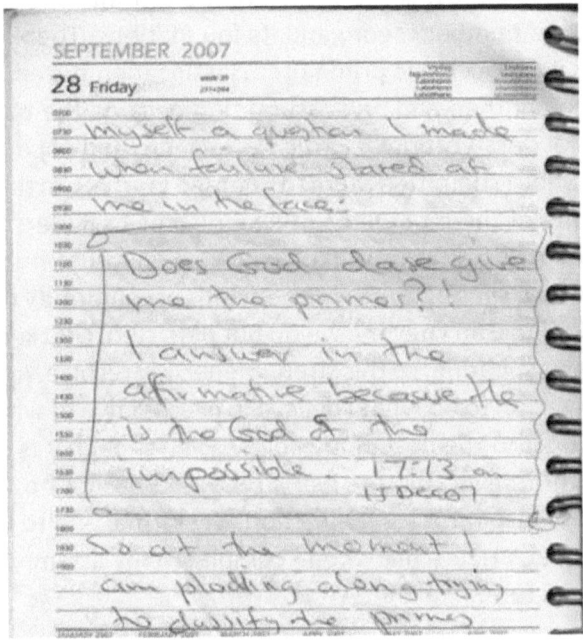

Fig-B. Inside my research diary

Secondly, being a Christian added another dimension. Mathematics is supposed to be a clinical subject having no connection to religious beliefs and perhaps even philosophical and doctrinal thought. As I developed my classification theory for the prime number, and noting a pattern that can be used to generate them, I was excited. But as normally happens in expository mathematics, you get great results from great disappointments. So my theory crash-landed with a heavy thud. I was really disappointed, but in my spirit, a simple prayer began to manifest. On the 15 December 2007 at 17:13, I made a very simple prayer as follows: "Does God dare give me the primes?! I answer in the affirmative because He is the God of the impossible. "I had no qualification as to what I really meant by "giving me the primes," that is, whether it meant finding a way to predict the next prime or proving the twin-prime conjecture. I think my mind was mostly on the classification system and a way of generating primes using patterns rather than complex mathematical formulas for a next prime. It was a daring prayer considering that I am a novice in issues of prime numbers—I lacked all the sophistication that is currently existing in the mathematical world.

My first record on the research into prime number was made on the 4 November 2007, at 07:07 a.m. I abandoned my key research on revolutions in mathematics for almost three years and delved into research activity on advances in prime number theory and its foundations. Then I worked on my patterns until I developed the prime root classification system. However, one discovery led to another, and one assumption became the base for the next leap until several classification systems were made. And of course, my papers were rejected when I submitted them to refereed journals. In the mathematical world, this is one of the most difficult experiences to understand because it is not how good or original you think your idea is; rather it is how relevant, contextual, and how satisfactory to the demands of conformity that counts. And I am also thankful to God I never gave up but persisted. God kept on giving me inspiration after inspiration, so this strengthened my resolve. And I came to the position of conviction to do what I believe in irrespective of what others judge it to be and to do it to the best of my ability. Even though it is not common in scientific thought to give God the glory, I have no hesitation whatsoever. He was my source of inspiration and direction. My formal end to the project was on 19 July 2010, at 11:22 a.m., when I completed my attempt of proving the twin-prime conjecture.

From an experiential perspective (or in the context of experiential mathematics), I would like to share some insights that I gained as a result of this project because as I said, it was also a form of personal experiment on mathematical development. In the context of constructing axioms, I noted the following:

   a. Discovery is not a rational process. [4 November 2007, 10:05 a.m.]
   b. The growth of knowledge based on discovery is a rational process. [4 November 2007, 10:35 a.m.]

The insight here is that often we assume that discovery is rational, but rather it is the growth of knowledge based on the discovery that is rational. The irrationality of what led to the discovery is what we normally don't display to the public, but it gives a false impression to those who may otherwise benefit from experiential mathematics as to how discovery comes about. I threw away a lot of my assumptions when I was focusing on publishing in a journal because of the emphasis of rationality and continuity from existing knowledge. But when I decided that I want to develop the idea independent of the set standard of journals, my creativity for discovery became free to explore and delve wider and deeper.

   c. There is a thought paradigm for any given knowledge area. A thought paradigm describes accepted current thinking in terms of meaning and interpretation in order to solve problems, create, and apply knowledge.
   d. The thought paradigm in which we exist depends on the dominant perceptions, where such perceptions are determined by the knowledge, classification codes, relationships, and associations in which we practice. [22 December 2007, 11:20 a.m.]
   e. Meaning is a key ingredient of creativity, and interpretation is a key factor for connectivity of knowledge and its development. [26 July 2010, 11:46 a.m.]

The great challenge of expository mathematics is that it tries to look for a new angle or explanation to existing problems, or it tries to create a new paradigm of thought with regard to given solutions or problems. Therefore, one has to be constantly aware of the existing paradigm of thought so that the point of departure is well-defined and justified. This is also a good mechanism for checking oneself in your conceptual development—what have others said, and what am I saying, and what are the major points of departure?

    f.   A theorem is formed in the context of a conceptual framework. [26 January 2008, 8:55 a.m.]

Normally, a theorem is a statement that is formed with the aim of providing a proof. When one wants to state an observation arising from certain conclusions, then a conjecture may be formed. My thinking is that one of the possible outcomes of an expository research in mathematics may be a theorem, provided it is formed in the context of a conceptual framework. Such a theorem is not just based on observations, which would limit it to a conjecture, but relies on mathematical evidence existing in the conceptual framework. This evidence may not necessarily form a proof but should provide ground for a theorem to be developed.

    g.   Delimitation is a powerful conceptual tool to clarify an idea, formulate a theory, and for the development of theoretical context. [27 January 2008, 09:52 a.m.]

This is a skill that one must attain in expository mathematics in order to have an ability to discern the value of ideas, their strength and capability to be developed further into strong concepts and models of thought. It reduces conceptual error and develops the context needed for the theory to have credibility.

    h.   Logic does not need much of imagination because it depends on already approved constructs based on existing premises. It relies more on the level of intelligence and analytic ability to produce result. However, it is also a powerful tool to validate the fruit of one's imagination. [26 July 2010, 16:33]

That is, at the end of the day, mathematics is mathematics whether expository or not. Whilst expository mathematics may have a lot of influence from the doctrinal context, as this is necessary to formulate meaning and interpretation, the final test must be able to withstand the demands of logical structure. Having gone out of the box because of imagination, one must be able to relate back to what happens in the box, and seeking logic always provides a framework for conceptual continuity.

The above are some of the insights that I picked up as a consequence of the research. The fact that I am passing them on does not mean that I was perfect in applying them, but they were very helpful in balancing my approach in developing the theory of prime number classification.

As to my major conclusion in regard to Crowe's laws, since I was conducting an experiment into how I think and form meaning, I made the following observations:

a. Mathematics is experiential, hence the term experiential mathematics. That is, it is not only connected to human thought, it is also connected to the human being. The experiential aspect of mathematics may be mostly evident in the creative process.
b. By bringing ideas into existence, definitions provide an opportunity for revolutions to occur *in* mathematics.
c. Proofs and axiomatic constructs provide a framework and methodology for mathematics to grow in a cumulative manner.
d. Mathematics is doctrinal. Ideas are born as a consequence of the framework of the mind-set, and as such doctrine acts as a filter to what imagination can present as an alternative.
e. Mathematics is logical. This is the essence of mathematics and forms the most prominent part of its presentation because of the need for communication of thoughts, ideas, and proofs.
f. Mathematics is intuitive. Intuition, which can be described as a mathematical sense that must be developed with effort, is the bridge builder between logical constructs. It is influenced by imagination, doctrine, ideas, and experience.
g. Expository mathematics can be outright revolutionary, but the growth of cumulative mathematics consists of paradigms of transition.
h. Expository mathematics eventually becomes part of cumulative mathematics when its doctrine and tradition are absorbed into current paradigms.
i. The emphasis of cumulative mathematics versus expository mathematics is summarized by the table below.

Expository Mathematics	Cumulative Mathematics
Depends on new ideas of thought as a premise for investigation and research.	Depends on existing research programs, tradition, and values as a premise for continuity and reference.
Starts with unbounded ideas, seeks new doctrine and methodologies to establish new paradigms	Starts with bounded ideas, using existing doctrine and methodologies to reinforce current paradigms
Consistency of idea-application, a critical value for discovery. The discovery process is unconditional.	Reference to existing mathematics is critical premise for acceptance of any idea. Discovery is contextual.
Intuition, imagination, discussion, ability to develop constructs of thought, development of ideas to concepts and to models are key tools.	Reason, logical frameworks, proofs, power of contextualization of thought to solve existing problems in current paradigm are key tools.

Table A. Difference in emphasis between expository and cumulative mathematics.

One of the objectives of this book is to *create a forum of discussion for those who are interested in the development of expository mathematics.* My common understanding of expository mathematics is that it is

> a subject of mathematical research on an idea-to-concept level in order to establish new structures and methodologies or framework of thought, where the information contained and derived sheds light and depth of meaning or interpretation to a current paradigm of knowledge.

Hence, even though the basis for mathematics is proof of argument, the emphasis of expository mathematics is investigation and idea building and, consequently, theoretical development. This book is presenting a theory; therefore, its emphasis is research and idea building. The results of expository mathematics are richer mathematical discussion emanating from new, improved, or expanded

- concepts,
- foundational definitions,

- mathematical structures,
- and methodologies.

Consequently, the other objective of the book is that of *providing an alternative approach to studying the properties and distributive behavior of prime numbers.* This explains the emphasis on descriptive methodology in contrast to analytic techniques. The idea is to illustrate that a lot can be done on a simpler level to conceptualize and develop ideas before they are crunched by other more powerful mathematical tools. The basis for the approach is made from the observation that in terms of underlying behaviors, numbers can be geometric or arithmetic in nature. The geometric character of numbers tends to be based on a descriptive framework as seen in the definition of a Cartesian space or the complex number system. Therefore, spatial concepts dominate the construction of the general theory of prime number classification.

Research Concepts

The book, consisting of several sections, is a comprehensive research that proposes four different methods of classifying prime numbers and the consequent application of the classification systems.

1. **The Prime Root Classification.** This assumes that every prime number has fundamental roots that describe its constitution. This produces a finite number of classes c such that $c < 8$. Prime numbers are also said to have prime signatures. A possible application of prime root theory is random event analysis.
2. **The Positional Classification.** This assumes that a prime number exists in a two-dimensional space with a unique coordinate. The classification provides prime-number dimensionalisation and derives the result that all the means of n groups of prime numbers always form a straight line. The classification also defines prime number count functions for a given range.
3. **The Delta Classification.** This classification is based on prime gaps and derives the important concept of a prime family. One of its results is that in terms of rate of change of a given gap, the prime gaps accelerate positively, negatively, or are steady. For example, gap 2 accelerates positively, gap 24 is steady, and gap 42 has a negative gap acceleration. Therefore, this classification is useful for analyzing gap behavior characteristics of prime numbers.
4. **The Gap Theory Classification.** Using prime families and prime gap axioms through this classification, one can demonstrate that there is a gap base system to describe all prime gaps, these being the pattern gap "2," gap "4," and gap "6." All other gaps of prime numbers can be described in distribution and behavior in terms of these three. Prime numbers can be precisely classified according to the gap base system.

Therefore, each classification system is regarded as technique for studying particular properties of the prime numbers. The last classification system is then used to research into how primes are generated, where this derives a form of sieve methodology.

Further research is developed along the following lines:

5. **The Gap Theory of Prime Number Generation.** This is an application of the classification theory to define how prime numbers are generated. The aim is not just to generate the prime numbers in a sequential manner but rather in the context of gap theory. That is, all prime numbers are generated in terms of gap base 2, gap base 4, and gap base 6 only using a special type of "simultaneous equation." Consequently, prime number-generation axioms are defined that allow the theory to demonstrate that all prime numbers are linearly related, thus explaining their regularity. The theory also defines the random function for the prime numbers. Hence, the theory explains the regularity feature of the prime numbers (where it comes from) and the source of the random characteristic.
6. **The Prime Number Generation Condition.** This condition formalizes the use of the expression $6a \pm 1$ as a methodology for generating prime numbers. On the basis of this approach, two equations are defined that derive prime numbers according to Gap Base 2 and Gap Base 4.
7. **The Sieve Theory of Prime Number Generation.** This defines an algebraic sieve that locates all numbers that are not prime on the sieve by means of an algebraic equation. The algebraic sieve is found to contain all the necessary information about prime numbers, including how gaps are formed and prime numbers distributed. It also explains how gaps actually expand and what causes them to expand at a particular point in the progression. For example, if you wonder why 5 and 7 are the only divisors necessary to sieve out primes less than one hundred, the algebraic sieve offers an exact explanation.
8. **The Primality Condition.** This is an algorithm using inverse operations in the algebraic sieve in order to define a condition that establishes whether a number is prime or not. It is probably the simplest and most accurate test ever designed. It consists of the P1-test (exclusion) that is also a composite test, the P1-test (inclusion) and the P2-test that test for primality. The P2-test is a probability test. The structure of the sieve

THE THEORY OF PRIME NUMBER CLASSIFICATION

creates a certainty for prime number occurrence. The primality test is also use to study patterns of composites, primes and pseudoprimes.

9. **The Law of Prime Numbers**. This is a theoretical approach that states that all prime numbers are generated numbers on the basis of the set $\{1, 5, 7, 11, 13, 17, 19, 23, 29\}$. It shows how the set of pseudoprimes are derived from this fundamental set notion and, consequently, defining prime number spaces. Also, the product of consecutive prime numbers or consecutive pseudoprimes or a combination of these in gap base theory lie on the prime number curve $36x^2 + 108x + 77$. The gap between all of these is always four.

10. **The Algebraic Prime Space**. Applying the law of primes derives the probability space for prime numbers, which then leads to the concept of an algebraic prime space. This space gives the density of prime numbers using the P-index, and is also used to plot the change in the prime number count.

11. **The Twin-Prime Conjecture**. Since the sieve contains all the necessary information about prime numbers, the logical path was then to find out whether it would be of assistance in creating a framework for proving the twin-prime conjecture. Therefore, an attempt is also made in that regard.

12. **The universal number distribution theorem.** This is a general approach to how numbers are classified and distributed with respect to each other in the context of primes, pseudoprimes and composites. The theorem justifies most of the approaches proposed by the classification theory. According to the theorem, all numbers are an algebraic expression.

Research Framework

The emphasis of each of the sections is as follows. Research questions were used as one of the tools to guide the intuitive process and to provide imagination a framework from which to operate.

Section One
Research Question

Natural numbers can be expressed as factors of prime numbers. Therefore, prime numbers are regarded as the building blocks of natural numbers (fundamental theorem of arithmetic). Hence,

> *What is the building block of prime numbers, or how are they constituted?*

Section Two
Research Question

Is it possible to define a concept of a prime space that is two-dimensional as a tool for distribution analysis?

RESEARCH FRAMEWORK

Section Three
Research Question

Is it possible to do a descriptive analysis of primes in terms of gaps influencing the prime distribution rather than primes influencing the gap distribution?

Section Four
Research Question

What is the influence of the prime gap on predicting the position of a given prime number in the prime space?

Section Five
Research Question

How does prime number classification assist in understanding prime number generation?

Section Six
Research Question

How can the sieve provide a framework for primality testing?

Is it possible to use the algebraic sieve to prove the twin-prime conjecture?

Section One

Root Classification of Prime Numbers

Overview

It is a common observation that prime numbers are random, but at the same time, they exhibit striking suggestive patterns. Through a theory of prime number decomposition where it is proposed that any prime number has either a prime root, an odd root, or an even root, a classification system in terms of roots of primes is developed.

Objective:

To develop a classification system of prime in terms of their basic constitution.

Research Question

Natural numbers can be expressed as factors of prime numbers. Therefore, prime numbers are regarded as the building blocks of natural numbers (fundamental theorem of arithmetic). Hence

What is the building block of prime numbers, or how are they constituted?

1 Introduction—Existing Classification

A prime number is any number that is positive and divides by one and itself only. This type of number is of special interest because it has been found to be the building block of all other natural numbers through the multiplicative process. The fundamental theorem of arithmetic asserts that all natural numbers that are not prime can be expressed uniquely as a product of primes. For example, 65 = 5*13, a product of two prime numbers. The interesting question, therefore, is

> *What is the building block of prime numbers, or how are they constituted?*

This question naturally makes one to investigate patterns of prime numbers in order to discover if there is any particular structure by which they can be defined. Currently, patterns have been established along the following lines:

a. The pattern can be arithmetic in nature. For example, there are Fermat primes in honor of Pierre de Fermat, and there are Mersenne primes in honor of the monk Marin Mersenne. The format for Fermat primes is $Fn = 2^{2^n} + 1$, and primes in this group are limited as none have been found beyond $n = 4$. However, there are more of Mersenne primes that are defined by Mn = 2^p - 1, where p is a prime number. There are simpler product patterns that defined primes such as $p = (6n + 1)$ and $p = (6n - 1)$.
b. The pattern can be described in a geometric context, such as being symmetric, repeating digits.

Other ways of looking at primes involve looking at properties of the primes in terms of their digits, such as being truncatable or being reversible. Such methods lead to forming another prime and, as such, cannot be said to look for an underlying pattern. They are looking for manipulative pattern attributes of the primes and common descriptive formations that are limited to a few primes. Another approach is to model primes according to their distribution patterns with the aim of discovering an underlying structure to their sequence. This proves daunting challenges since such a distribution is random in nature.

Whatever the case, these do not define an underlying structure that explains how prime numbers are constituted. As a result, there are many types of primes, but all these are distinct groupings that have no commonality. However, though they cannot provide a holistic classification framework, they provide interesting research into primes and their curious properties.

Therefore, the aim here is to identify common structural attributes that can be used to classify any prime number. Classification is an important tool in the sense that it provides a basis of describing underlying patterns that are common to all primes. That is, though prime numbers are known to be random in nature, there could be certain specific patterns that may be observed in their randomness to suggest classification. An unpublished paper by Garavagalia and Garavagalia[1] define an algorithm that classifies primes according to location using a concept of ordered array on integers. Eismann[2] developed a concept of decomposition of natural numbers and applied it to the classification of prime numbers by assuming that a prime has the characteristic of weight × level + jump. A notable classification of primes is made by Paul Erdös and John Selfridge.[3]

It appears that two accepted ways of classifying prime numbers are using class $n+$ and class $n-$ as suggested by the work of Paul Erdős and John Selfridge. They devised a methodology where the class $n+$ of a prime number p involves looking at the largest prime factor of $p+1$. That is, if the largest prime factor is 2 or 3, then p is of *class 1+*. Then if the largest prime factor is another prime q, then the class $n+$ of p is one more than the class $n+$ of q. Similarly, the class $n-$ is almost the same as class $n+$, except that the factorization of $p-1$ is determines the methodology.

2 Basis of Theoretical Research

There is a famous quotation made by Don Zagier in a 1975 lecture that reads as follows:

> *There are two facts about the distribution of prime numbers of which I hope to convince you so overwhelmingly that they will be permanently engraved in your hearts. The first is that, despite their simple definition and role as the building blocks of the natural numbers, the prime numbers grow like weeds among the natural numbers, seeming to obey no other law than that of chance, and nobody can predict where the next one will sprout. The second fact is even more astonishing, for it states just the opposite: that the prime numbers exhibit stunning regularity, that there are laws governing their behavior, and that they obey these laws with almost military precision.* (Havil 2003, p. 171)

As you know, a weed appears anywhere unexpectedly, and prime numbers exhibit such a behavior in their distributive pattern. The comparison to a weed is extremely apt since a weed still obeys the laws of growth though its next location cannot be determined. Two problems are highlighted by this quote and analogy:

a. Firstly, how we consider sequencing of primes as a limitation to understanding how the next prime will occur. We observe that there is no predictability because our sequencing is based on a prime n and a prime (n+1). Classification, for example, does not focus on the concept of the next prime, rather the issue is structure. Sequencing is a notion of counting rather than an expression of foundational blocks

for formation. That is, there are two characteristics when one considers the prime number.
b. Secondly, how are prime numbers constituted if there is a definite and precise manner of describing their behavior? If other composite numbers are expressed in terms of the primes, logically one must assume that there is a methodology of formation for the primes. Formulas generally fail to be universal in establishing this methodology.

Rather than focus on using only formulas to determine how prime numbers are formed, a non-arithmetic approach was considered. Hence, certain observed occurrences informed the basis of the research work. Consider for example the prime numbers 2711 and 2749. We then observe that the number 2711 can be decomposed in three prime primes, that is 2, 7, and 11. Such primes are then called roots of 2711. This is designated [prime, prime, prime] where the notation indicates the number of prime roots in terms of structure. The same prime number can also be decomposed as roots 271 and 1, which is [prime, odd]. In order to resolve this and create a unique solution, the priority rule is developed. Similarly, 2749 can be decomposed into two primes, that is 2 and 7; hence, the remaining number 49 is called an odd root. On the right, the remaining number will always be an odd root if it is not prime since all prime numbers are odd.

Secondly, for example, consider the prime number 181 as illustrated for primes less than 20000 in table 1. There is a distinct left and right pattern in identifying a prime root, provided the prime root search starts either from the left or right. Note that a prime number such as 11813 with 181 in the middle does not define a left and right operation. In this case, the prime roots are identified as 1181 and 3. However, the 181 in 11813 is called a prime signature, and the number 11813 is called the first prime signature of 181.

Left Root 181	Right Root 181
*181*1, *181*19, *181*21, *181*27,	1181, 3181
*181*31, *181*33, *181*43,	9*181*, 10*181*, 19*181*
*181*49, *181*69, *181*81, *181*91, *181*99	

Table 1. The left and right behavior of prime roots.

An interesting case is the prime 18181, the prime root can either be a left or right root. The correct procedure to find roots is therefore determined additionally by use of the priority rule in order to derive unique results.

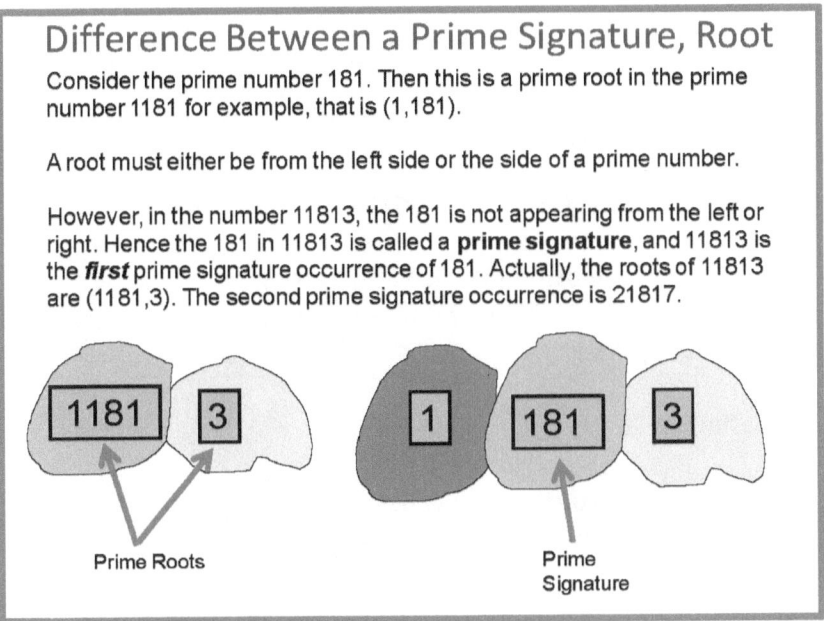

Difference Between a Prime Signature, Root

Consider the prime number 181. Then this is a prime root in the prime number 1181 for example, that is (1,181).

A root must either be from the left side or the side of a prime number.

However, in the number 11813, the 181 is not appearing from the left or right. Hence the 181 in 11813 is called a **prime signature**, and 11813 is the **first** prime signature occurrence of 181. Actually, the roots of 11813 are (1181,3). The second prime signature occurrence is 21817.

Figure 1. Identifying the prime signature and root.

3 Prime Number Decomposition

The concept of decomposition is based on looking for components within a prime number that form distinct number patterns that can be regarded as being roots of the primes.

> **Definition 1**
>
> *A positional decomposition of a prime number is a breakdown of the prime such that distinct whole numbers in sequence can be derived.*

For example, a decomposition of 2711 is denoted as (2, 7, 11), where 2, 7, 11 are distinct whole numbers in sequence such that in putting together the roots, that is on performing a composition, gives us the prime number again. It is for this reason that every decomposition must form a proper number so that the converse operation is also true. However, for a prime such as 503, if we decompose this to 5, 03, then 03 = 3, and this would imply 53 when we compose the numbers again. Hence, 03 is not a root of the prime number and 503 is a nondecomposing prime. Specific rules apply for a decomposition such that we are able to define prime number classification. For example, for the prime 55021, then (5, 5, 021) is an incorrect decomposition. According to the rules that will be explained, 55021 gives (5,5021), but this would be different for the prime 55061 because this would have the root decomposition (550, 61).

> **Axiom 1**
>
> *A root r of a prime p is any number $r \leq p$ that can be identified in p that is either prime, even or odd in that priority.*

THE THEORY OF PRIME NUMBER CLASSIFICATION

If r is a prime number, then it is called a prime root. Similarly, r can be even or odd, forming even or odd roots respectively.

Axiom 2

In a prime number decomposition, prime roots exist in a sequence either from the left or right of a given prime number.

When prime roots are derived in a sequential manner, this is referred to as a root sequence. All prime roots are also prime numbers, and a prime root is always smaller than the prime number in which it is found, hence the use of the word "roots." According to axiom 2, the roots are identified uniquely in sequential format either from the left or right. We also note through axiom 2 that if r_{i-1} and r_{i+1} are prime, then r_i is prime. Hence, it is not possible according to the axiom to have a format [prime, even, prime], [odd, prime, odd], or [even, prime, odd] because the prime roots are not consecutive or the prime root is not the first from either the left or right.

Definition 2

In a prime number decomposition operation ϕ,

a. *A right operation denoted as ϕ_R defines a biggest prime root starting a sequence from the right of a given prime.*

b. *A left operation denoted as ϕ_L defines a biggest prime root starting a sequence from the left of a given prime.*

Definition 3

Assume that for any given prime number $p = (a_1 \ldots a_n)$ where a_i represents a digit, then the prime has m roots $(r_1, \ldots, r_m)$, such that,

a. $(r_1, \ldots, r_m) = (s_1, \ldots, s_m)$, *for a left operation*

b. $(r_1, \ldots, r_m) = (s_m, \ldots, s_1)$, *for a right operation*

3 PRIME NUMBER DECOMPOSITION

Therefore,

$$\phi p = \phi(a_1 \ldots a_n) = (r_1, \ldots, r_m) \quad (1)$$

$$\phi_L p = (s_1, \ldots, s_m) \quad (2)$$

$$\phi_R p = (s_m, \ldots, s_1) \quad (3)$$

Consequently, we observe that a given prime root can occur on the right of a prime number or on the left of a prime number and $(r_1, \ldots, r_m)$ is called a sequential root decomposition of the prime number.

Theorem 1

Whilst prime number occurrence is random, there exists a specific underlying structure to prime number formation.

Generally, three types of roots are recognized—even, odd, and prime. According to the theorem, any decomposing prime number must have either one of these types of roots. The proof of the theorem lies in showing that a classification system represents such a structure.

Proof

For any prime number p composed of n digits, then let $p = (a_1 \ldots a_n)$ where a_i represents a digit. Let any number of such digits a_i be combined together without change of position to form distinct but separate numbers, then the numbers so formed will either be an even number or an odd number, and some of the odd numbers may also be prime. That is, $a_1 \ldots a_i = r_1$, $a_{i+1} \ldots a_{i+j} = r_2$ and generally $a_x \ldots a_y = r_i$ where $x < y, y \leq n$. This defines a positional decomposition where we assume the sequence of roots $(r_1, \ldots, r_m)$ for the prime p provided r_i is a distinct number.

Hence, according to axiom 1, let there be a positional decomposition of p. Then axiom 2 is applied to the decomposition as follows. Let there be a defined root sequence from either the right of left of a given prime, then from equation 1, we consider the roots $(r_1, \ldots, r_i, \ldots, r_m)$. Hence,

THE THEORY OF PRIME NUMBER CLASSIFICATION

a. If r_i is a prime root from a left operation, then r_1 to r_i consists of prime roots only. Consequently, p is called a standard prime. Hence, r_{i+1} to r_m will be an odd number provided $r_{i+1} \neq 0$. Therefore, there can only be one odd root on the right of a prime number p, that is, $r_{i+1} = r_m$.

b. If r_i is a prime root from a right operation, then r_i to r_m consists of prime roots only. Consequently p is called a standard prime. Hence r_{i-1} to r_1 will be an odd number or an even number. Therefore, there can only be one odd or even root on the left of a prime number. That is, $r_1 = r_{i-1}$.

c. Let r_1 be a prime root and r_m a prime root, then r_i is also a prime root since roots are in sequence. Hence, the prime number consists of no even or odd roots and is called pristine.

d. If r_i is a prime root from a left operation, and let $r_{i+1} = 0$, then $r_{i+1}, \ldots, r_m$ does not define a distinct number that forms a root since $r_{i+2} = r_m \neq r_{i+1}$ to define a proper left operation. Hence, this defines a nondecomposing prime. Since the defined root sequence is left, then such primes can always be identified from a left operation.

e. Let $r_1 = x$ be an odd root then define $r_2, \ldots, r_m$ as y where this either consists of prime roots or composes an odd number. If there is no prime root, then y is an odd number since all prime numbers are odd. Hence, the prime number consists of odd numbers and is defined as a rare prime. If there is a prime root in y, then r_2 must be a prime root according to axiom 1, that is, for a root sequence to exist. Hence, we assume no prime root in order to show that there exists a rare prime.

f. Let $r_1 = x$ be an even root then define $r_2, \ldots, r_m$ as y where this either consists of prime roots or composes an odd number. If there is no prime root, then y is an odd number since all prime numbers are odd. This defines a root unique prime, that is, a prime number that does not consist of prime roots but consists of even and odd roots only.

Therefore, all prime numbers have an underlying structure defined by ϕp.
∎

Generally, a root unique prime has no prime root. Therefore, it may be assumed that a rare prime is also a root unique prime since it does not consist of any prime roots; the only difference is that it does not have an even root. Also two is the only prime that has an even root only, and 3, 5, 7 consist of prime roots only; hence, they are example of pristine primes. The consequence of the theorem is that

1. if every prime number decomposition has a root, then a form of classification system can be developed;
2. if primes have a distinct structural pattern of formation, then this defines a methodology to study how prime numbers occur such that predictive distribution analysis is possible; and
3. if a prime number is defined by a prime root sequence, and the remaining part is always either an odd or even root, then odd and even roots are always in sequence to prime roots.

3.0.1 NONDECOMPOSING PRIMES

A nondecomposing prime has been defined as one that does not have a prime root sequence because of the occurrence of a zero as part of the roots. This zero is called the zero root of a nondecomposing prime. Hence, it is also evident that for a nondecomposing prime number, the general pattern of such is

1. $[prime]\ldots[prime][0][odd]$ since the only possibility on the right is an odd root, for example, 5209 is [prime][prime][0][odd] where 5 and 2 are prime roots and 9 the odd root.

2. $[prime]\ldots[prime][0][prime]\ldots[prime]$. That is, the zero is between a prime root sequence, for example,

   a. 503 is $[prime][0][prime]$, and
   b. 1201027 is $[prime][0][prime][prime]$, where 1201 is prime, then zero followed by 2 and 7.

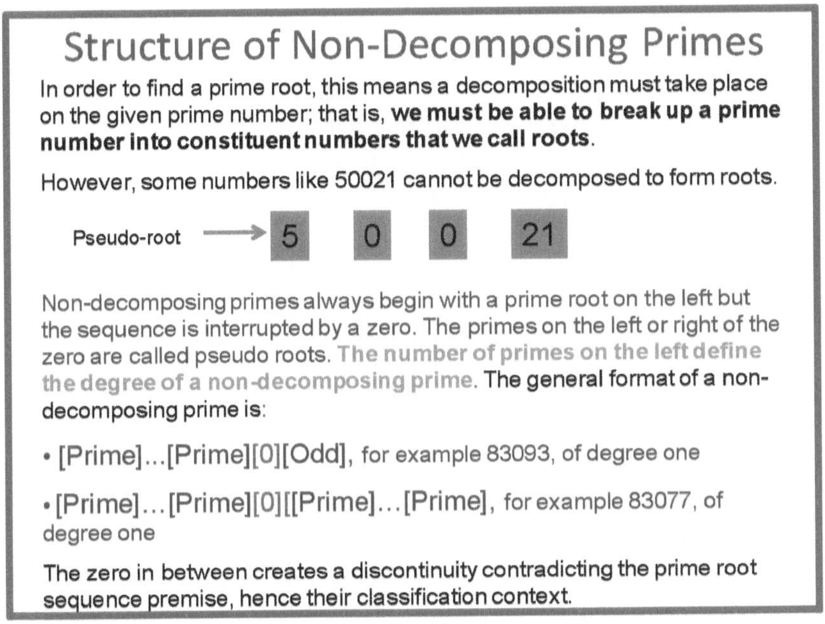

> ## Structure of Non-Decomposing Primes
>
> In order to find a prime root, this means a decomposition must take place on the given prime number; that is, **we must be able to break up a prime number into constituent numbers that we call roots**.
>
> However, some numbers like 50021 cannot be decomposed to form roots.
>
> Pseudo-root ⟶ | 5 | 0 | 0 | 21 |
>
> Non-decomposing primes always begin with a prime root on the left but the sequence is interrupted by a zero. The primes on the left or right of the zero are called pseudo roots. The number of primes on the left define the degree of a non-decomposing prime. The general format of a non-decomposing prime is:
>
> • [Prime]...[Prime][0][Odd], for example 83093, of degree one
>
> • [Prime]...[Prime][0][[Prime]...[Prime], for example 83077, of degree one
>
> The zero in between creates a discontinuity contradicting the prime root sequence premise, hence their classification context.

Figure 2. General rules for non-decomposing primes.

The zero root breaks the overall root sequence of the prime number, hence defining the non-decomposition characteristic. In such a prime, the prime root exists, but a whole number decomposition is not defined, so the roots are not recognized. The prime number may be said to consist of pseudo-roots.

Assuming then those non-decomposing primes have pseudo-roots, this implies we can also define a classification for them based on their particular characteristic since internally they have prime roots from a left operation.

Definition 4

A non-decomposing prime is of degree n, where n is the number of pseudo-roots before the zero root.

Therefore, 5209 is a non-decomposing prime number of degree two, while 1201027 is of degree one. The definition ignores any prime root occurrence after the zero root because the non-decomposition is defined on the basis of a left operation.

3 PRIME NUMBER DECOMPOSITION

3.0.2 THE CLASSIFICATION RULES

According to axiom 1, there exists a root r in p such that $r \leq p$, and axiom 2 says any prime roots must be in sequence. The two axioms establish the conceptual structure of a prime number, and theorem 1 has the consequence of deriving a classification framework on the basis of finding the roots of prime numbers. Hence, in order to develop unique solution of the roots, the priority rule is applied where the consequence of the rules is that,

a. it defines a relationship between the three types of roots of a prime number by providing a deterministic ranking;
b. it gives concise structure to defining the roots of any given prime number, otherwise the same primes would have different roots to construct them;
c. it gives us a specific way of determining an even root and, consequently, an odd root; and
d. without the priority rule, then there is a possibility that an infinitely large prime will have a relatively large number of prime roots. Therefore, this minimizes the number of roots a prime number can have.

Axiom 1 also defines a basis for the priority rule because it assumes that the order of priority in roots must be a prime root, an even root, and an odd root. The priority rule consists of a set of sub-rules defined as follows:

Definition 5

Priority Rule

1. In a given prime number p with roots $(r_1, \ldots, r_m)$, the prime root from the left or right operation must be the largest possible outcome.
2. After the first biggest prime root, the next biggest prime root is identified in sequence.
3. The next biggest prime root may be smaller or bigger than the previous prime root.
4. The bigger root either from the left or right determines whether we have a left or right operation respectively.

5. The prime root takes priority over the odd root.
6. The even root takes priority over the odd root, where the root is the largest possible even number from the left of the prime number.

In the case $\phi_L 3581 = (3, 5, 81)$, we note that the first prime root is smaller than the next prime root, and it is a left operation because prime root has priority over the odd root 81. The prime root 3 is regarded as the first biggest prime root from the left because 35 is not prime. The same applies for a prime such as $\phi_L 41357 = (41, 3, 5, 7)$. This is a pristine prime. It is always easier to look for a prime number first. For the prime number 9649, there is no prime root constructing this number either from the left or right, so according to the theorem, it is a unique prime number. Therefore, using the priority rule, it can be written as $\phi_L(9649) = (964, 9)$ because the even root has priority, and it has format [even, odd]. For the unique primes such as 19 and 9391, starting from the right or left, there is no even number. That is, the prime has the format [odd] and has no even root. Such a prime number is its own root and, as such, is a unique prime but is referred to as being rare as it has no even root.

Hypothesis 1

The priority rule constrains the total number of prime roots m for $(r_1, \ldots, r_m)$ to less than eight.

The number 8 is derived from observing the pattern of class occurrences through the use of a computer program. From the hypothesis, it does not imply that the larger the prime, the larger the number of roots as one would expect. The reason for the limited number of roots no matter how large a prime is based on the fact that we always look for the biggest prime from the left or right; therefore, as the primes become larger, the roots also become larger. That is, if p is a prime and $(a_1 \ldots a_n)$ so that each term represents a single digit, then $p = a_1 \cdot a_2 \ldots a_n \times 10^y$

a. the largest possible prime root will be less than or equal to
$a_2 \cdot a_3 \ldots a_n \times 10^{y-1}$
for a right operation, and $1 \leq a_2 \leq 9$.
b. the largest possible prime root will be less than or equal to
$a_1 \cdot a_2 \ldots a_{n-1} \times 10^{y-1}$

for a left operation, and $1 \leq a_1 \leq 9$.

Such a prime number if it is not unique, and no matter how large, has at least one prime root either from the left or right. The roots become bigger as the prime numbers also become bigger. Hence, we do not expect a prime number to consist of an infinite number of root terms. A simple example is the prime given by the following:

$$\phi_R(3576863126462165676 29137) = (3, 576863126462165676 29137)$$

It is extremely large but consists of two roots only. Without the priority rules, we could have $\phi_R(1777771) = (17, 7, 7, 7, 71)$, which would be five roots. However, by applying this rule, we have $\phi_L(1777771) = (1777, 7, 71)$, making a reduction from five roots to three. This is the constraining effect of the rule.

4 Classification of Primes

Conceptually, classification of primes may be regarded as having two basic forms: the intrinsic classification and the extrinsic classification. In the intrinsic classification, the actual structure of the prime determines its classification, whereas in the extrinsic classification, grouping is defined by use of indirect parameters. For example, in the Erdös-Selfridge classification of primes, we find the highest prime factor of $(p+1)$, where p is the prime number. We do not use the prime number itself but rather use the factor of $(p+1)$ to determine the grouping of p. In an intrinsic classification, we focus on the structure of p; hence, using prime roots is an intrinsic classification approach. An intrinsic approach also enables us to directly

Class	Subgroup	Root 1	Root 2	Root 3	Root 4
Class 1	Nondecomposing	None	None	None	None
	Unique:—Rare	Odd	None	None	None
	Unique	Even	Odd	None	None
Class 2	Standard-Pristine	Prime	None	None	None
	Standard	Odd	Prime	None	None
	Standard	Prime	Odd	None	None
	Standard	Even	Prime	None	None
Class 3	Standard-Pristine	Prime	Prime	None	None
	Standard	Odd	Prime	Prime	None
	Standard	Prime	Prime	Odd	None
	Standard	Even	Prime	Prime	None
Class 4	Standard-Pristine	Prime	Prime	Prime	None
	Standard	Odd	Prime	Prime	Prime

43

4 CLASSIFICATION OF PRIMES

	Standard		Prime	Prime	Prime	Odd
	Standard		Even	Prime	Prime	Prime

Table 2. Demonstrating the classification of primes using roots up to class 4.

compare properties of the primes because we are using the same internal characteristic to evaluate them.

Definition 6

For any prime p where $\phi p = (r_1, \ldots, r_m)$ with m roots, then the class c of a prime is defined by $c = (x + 1)$, where x is the number of prime roots, and $c \leq 7$.

For prime classes, $c \leq 7$ because of the hypothesis arising out of the priority rule. The class 1 set of primes is defined by $x = 0$; hence, the class is 1. This class consists of all primes that do not have a prime root. This is defined by non-decomposing primes, rare primes, and root unique primes. Similarly, class 2 are primes that are constructed from only one prime, class 3 primes are constructed from two primes, and so on as indicated in table 3.

Class	Primes	Classification
Class 1	$\phi_L(166, 9) = 1669$	[even, odd]
Class 2	$\phi_R(540, 7) = 5407$	[even, prime]
Class 3	$\phi\_L\,(3, 5, 81) = 3581$	[prime, prime]
Class 4	$\phi\_L\,(3, 2, 2, 1) = 3221$	[prime, prime, prime]

Table 3. Example of classification using prime roots.

Theorem 2

No root unique prime number is a root of another root unique prime number.

Proof. Assume that p_1 is a root unique prime number, and p_2 is a root of p_1. If $\phi p_1 = (r_1 \ldots r_m)$ then $p_2 = r_i$. Now if r_i is prime, then p_1 is not root unique; therefore, it must be either an even or an odd root. Hence, no root unique prime number is a root of another root unique prime number. [end proof]

THE THEORY OF PRIME NUMBER CLASSIFICATION

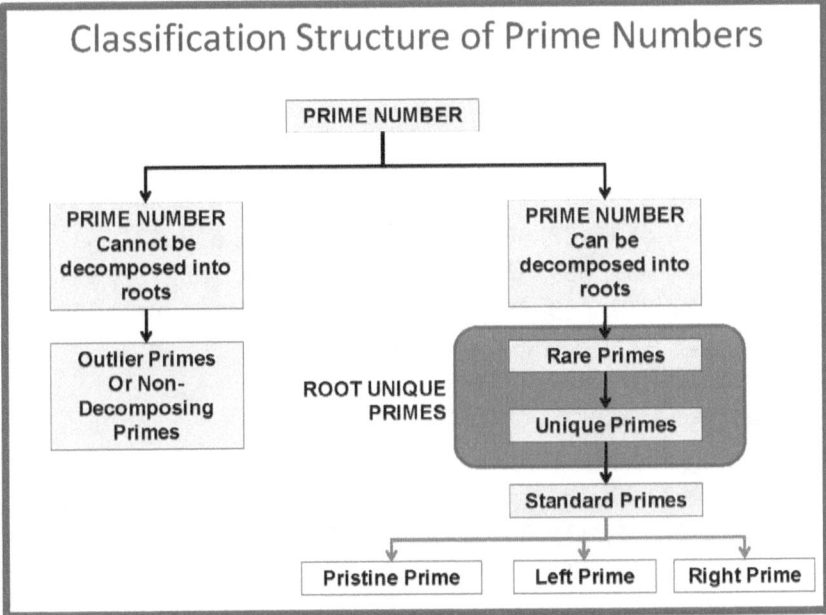

Figure 3. Using the property of roots to form classification tree.

However, unique primes are roots of other non-unique primes. Consider the unique prime 181. Though it is not defined by other primes, it defines other primes such as 1181, 1811, 3181, 9181, 10181. Alternatively, consider the unique prime 991. It is a root in the following primes 6991, 15991, 19991, 21991. Hence, unique primes are roots of other non-unique primes.

Example 1

The following is an example of a classification of primes less than one thousand.

Class 1 (no prime root)

> *Non-decomposing*: 503, 509, 701, 709 (all are of degree one)
>
> *Unique*: 41, 61, 89, 101, 109, 149, 181, 401, 409, 421, 449, 491, 499, 601, 691, 809, 821, 881
>
> *Rare*: 11, 19, 151, 991

4 CLASSIFICATION OF PRIMES

Class 2 (one prime root)

Pristine: 2, 3, 5, 7

Standard: 13, 17, 29, 31, 43, 47, 59, 67, 71, 79, 83, 97, 103, 107, 113, 131, 137, 139, 163, 167, 173, 179, 191, 197, 199, 239, 263, 269, 281, 307, 311, 349, 419, 431, 439, 443, 461, 463, 467, 479, 487, 563, 569, 587, 599, 607, 619, 631, 641, 643, 647, 653, 659, 661, 673, 683, 719, 739, 769, 787, 811, 823, 829, 839, 853, 859, 863, 883, 887, 907, 911, 919, 929, 937, 941, 947, 953, 967, 971, 983, 997

Class 3 (two prime roots)

Pristine: 23, 37, 53, 73, 193, 211, 223, 229, 233, 241, 271, 283, 293, 313, 317, 331, 337, 347, 353, 359, 367, 373, 379, 383, 389, 397, 433, 523, 541, 547, 571, 593, 613, 617, 677, 733, 743, 761, 773, 797, 977

Standard: 127, 157, 251, 457, 521, 593, 751, 827, 857, 877

Class 4 (three prime roots)

Pristine: 227, 257, 277, 557, 577, 727, 757

Class 5: None

The classification has a finite number of classes which are assumed to be less than or equal to seven. The first class 5 prime number is 3527 followed by 5527; both are pristine. The first class 6 prime is 27527 then 57527 followed by 77527, and all are pristine.

The table 4 shows the distribution of primes by class for one up to one hundred thousand in subgroups of 10000. Occurrences of primes at class 6 are not common, only three in the first one hundred thousand. The following definition is made with distribution properties of prime numbers.

4.1 Distribution Analysis

The classification is useful as a tool for distribution analysis tool of the prime numbers. Two main ways are considered under the classification system looking at the distribution in terms of class and in terms of range. The class distribution gives a distribution that reflects the structure of the prime numbers, whilst in range, the focus is the percentage distribution of the prime type. A range is said to be similar to another if both ranges demonstrate the same approximate distribution characteristics statistically.

4 CLASSIFICATION OF PRIMES

General Pattern of Class Distribution

This distribution is the number of primes less than 100 000, where each column is multiplied by 10^3, and their distribution by class. Generally, Class Two has the highest number of primes, followed by Class Three.

	1-10	10-20	20-30	30-40	40-50	50-60	60-70	70-80	80-90	90-100	Total	%Total
Class 1	142	72	51	28	76	50	58	53	84	59	673	7%
Class 2	600	536	411	340	485	382	610	316	485	580	4745	49%
Class 3	403	356	366	474	315	338	176	366	263	203	3260	34%
Class 4	82	66	137	96	50	131	34	144	42	36	818	9%
Class 5	2	3	17	20	4	22	0	22	2	1	93	1%
Class 6	0	0	1	0	0	1	0	1	0	0	3	0%
Total	1229	1033	983	958	930	924	878	902	876	879	9592	100%

Table 4. Showing the distribution of prime numbers per class

For each range, there is a similar general pattern of the prime distribution by class.

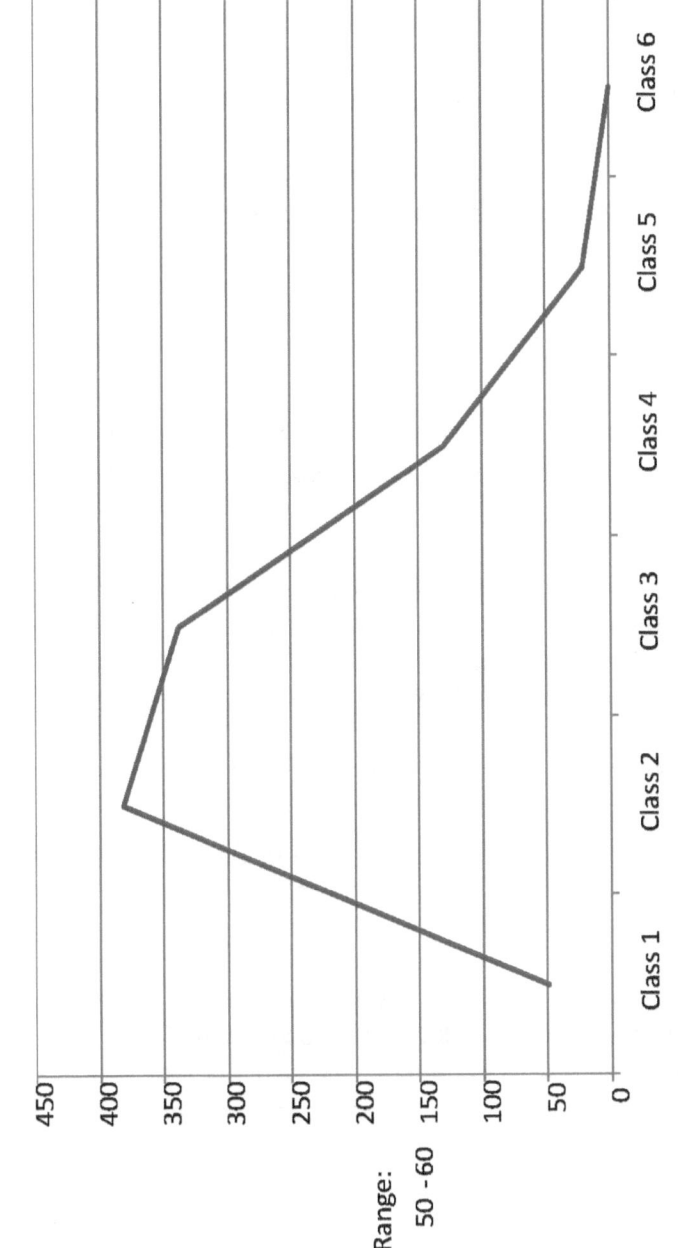

Figure 4. The prediction is that there are no classes beyond c = 7.

4 CLASSIFICATION OF PRIMES

Conjecture 1

The classification system gives us a predictive tool in terms of describing the type of primes we expect in a given range.

The prediction is based on using the range to describe the primes under study and the expected range in which we predict similar characteristics. It must be noted that two given ranges, these may have different prime distribution statistically, but what the conjecture highlights is the fact that there are ranges within the prime number spectrum that have similar statistical characteristics.

Definition 7

For a range $[n,m] \times 10^a$ where a is an integer, $n < m$ and $m \leq 9$, then this is a similar range to $[n,m] \times 10^b$ where b is an integer and $a < b$.

The relationship between similar ranges is that they are in powers of ten, and therefore the definition implies that prime number distribution patterns are sensitive to decimal position. If that is the case, then the properties of the classification system should be able to reveal such a trend.

4.2 PROPERTIES OF THE CLASSIFICATION SYSTEM

The classification is descriptive in terms of pattern analysis and systematic in terms of characteristics that describe the distribution of the primes. The following will be observed:

For *class 1*, rare primes begin with 1 and 9 on the left. Consider primes between 1000 and 10000, then if we find no rare primes from 1000 to 2000, the next place to look at is from 9000 to 10000. In fact, the only rare primes are 9391 and 9551. Similarly, we can say there will be no rare prime from 20000 to 90000; the only rare primes to hundred thousand are 915991 and 95539. The same argument would apply for two million to nine million for example. This is a predictive attribute from the classification.

THE THEORY OF PRIME NUMBER CLASSIFICATION

For *class 1*, unique primes begin with 1, 4, 6, and 8 on the left. For example, there will be no unique primes between 5000 and 6000 or 50000 and 60000. This is a predictive attribute that depends on the chosen range.

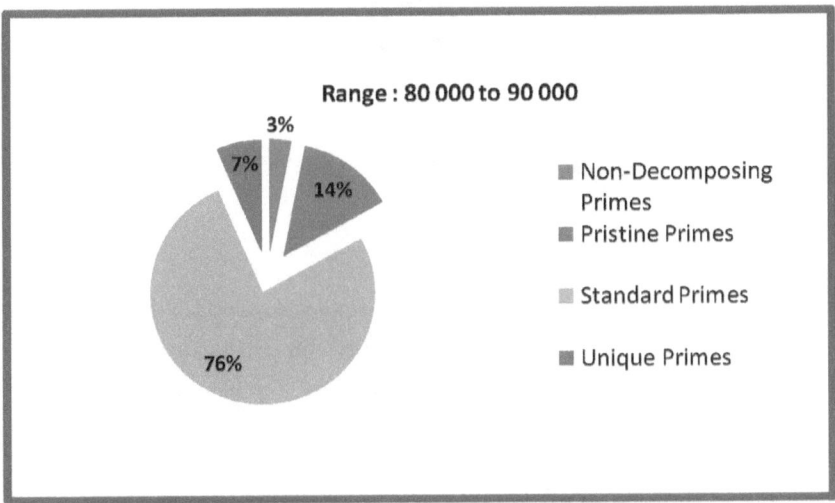

Figure 5. Showing distribution by prime type for a given range.

For *class 1*, non-decomposing primes begin with prime numbers on the left. The non-decomposing prime occurrence is influenced by the prime number sequence starting with 2, then $3, 5, 7, 11, 13, \ldots, \infty$. The sequence may be skipped depending on the definition of the range. For example, from one to one thousand, the nondecomposing primes start from the prime number 5, skipping 2 and 3, where the first prime is 503. Each prime number is said to form a cluster of non-decomposing primes. Because of the sequence in prime numbers, we can predict where to look for the next cluster of non-decomposing primes. The occurrence of such clusters is partially a function of the prime gap since the next prime defines the next cluster.

4 CLASSIFICATION OF PRIMES

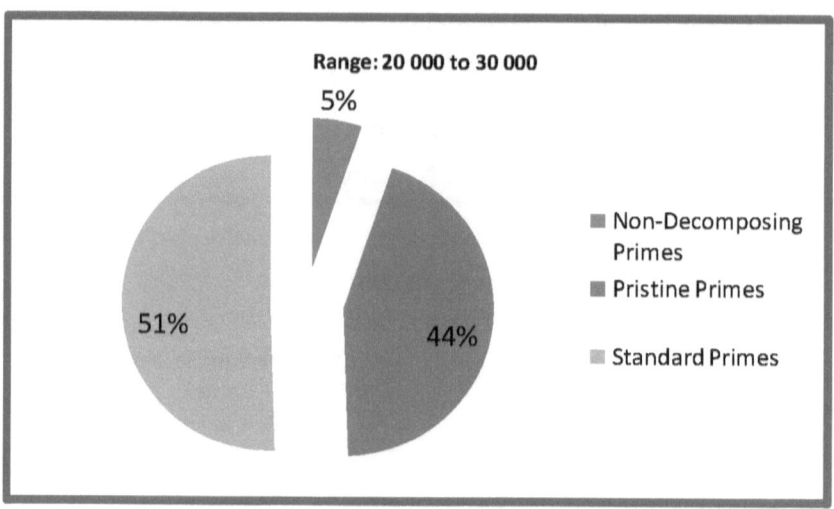

Figure 6. Showing distribution by prime type for a given range.

Class 2, they are all standard primes with the exception of 2, 3, 5, 7 since these are pristine. The primes in this class begin with any number and have prime roots either on the left or right. This explains why they are the most populous. Statistically, the standard prime is the most dominant group of primes, hence the use of the word "standard." This is a non-predictive group of primes in terms of using range.

Class 3 tends to have the highest occurrence of pristine primes for a reasonably large range, though this can differ depending on chosen range, and standard prime dominates.

Class 4 onwards, there are fewer primes comparatively, and there is also a tendency of a strong dominance toward pristine primes.

Class 5-7, these are predominantly pristine primes, and they start with a 2, 3, 5, and 7 on the left or right of the prime number. Their count is comparatively insignificant for a large range.

Consequently, using this information, one may make a guess as to the likely location occurrence in range of a given type of prime for a given class. Taken generally, the distribution of the primes tends toward a bell—or spike-shaped distribution for a reasonably large range peaking at class 2. This gives us a generalized statistical model of prime number distribution.

The importance of classification is that it illustrates the fact that even though prime numbers are random in nature, they have a specific structure. The fact that the number of classes is not infinite but is less than or equal to seven creates a fairly strong hypothesis that, although random, primes have a distinct pattern and structure in their formation. Classification provides the following:

1. **Structure**. The classification technique uses structure of the prime number to determine the class. The structure is defined by the notion of prime number roots. Hence, the classification is able to define a specific structure for very prime number.
2. **Pattern**. A classification pattern is developed through the use of a table. This classification pattern shows the distribution of the prime numbers by class and subgroups within the class. The distribution pattern has the form of a skewed-bell shape or spike.

Type. Occurrence by type in prime number distribution can be looked at from two perspectives: that of a given range and that of prime class. The classification demonstrates the fact that the distribution of primes is a factor of the chosen range. Firstly, certain primes are more likely to occur in a given range compared to other types. That is, the range 20000-30000 will have a different distribution pattern to 80000-90000. This implies that distribution is sensitive to the decimal position of the prime. Secondly, for a range $[n,m] \times 10^a$ and a similar distribution range $[n,m] \times 10^b$, the statistical description of the range is similar. For example, we expect the range 2000 to 3000 to have a similar statistical distribution to 20000 to 30000. Thirdly, prime types dominate according to a particular class, such as pristine primes in class 3 onward and standard primes in class 2.

4 CLASSIFICATION OF PRIMES

Comparative Analysis:
Pristine and Standard Distribution

This distribution is the number of primes less than 100 000, and their distribution by class according to Pristine and Standard categories.

A Pristine **B** Standard

	1-10		10-20		20-30		30-40		40-50		50-60		60-70		70-80		80-90		90-100		Total
	A	B	A	B	A	B	A	B	A	B	A	B	A	B	A	B	A	B	A	B	
class 2	4	596	0	536	0	411	0	340	0	485	0	382	0	610	0	316	0	485	0	580	4745
class 3	296	107	168	188	289	77	426	48	130	185	273	65	82	94	291	75	99	164	82	121	3260
class 4	78	4	45	21	129	8	94	2	30	20	122	9	19	15	136	8	21	21	14	22	818
class 5	2	0	2	1	16	1	19	1	4	0	20	2	0	0	21	1	2	0	1	0	93
class 6	0	0	0	0	1	0	0	0	0	0	1	0	0	0	1	0	0	0	0	0	3
Total	380	707	215	746	435	497	539	391	164	690	416	458	101	719	449	400	122	670	97	723	8919

Figure 7. Pristine and standard type in prime number distribution.

General Pattern of Class Distribution

This distribution of the primes is also sensitive to the first digit of the prime number. This sensitivity is different for each class. By sensitivity it is meant that certain classes will yield more primes because of the range defined by the first digit of the prime numbers. This sensitivity allows a contextual prediction to be made about occurrences of primes in terms of patterns. Polynomial trendlines of degree five show the behaviour difference for Class Two and Three.

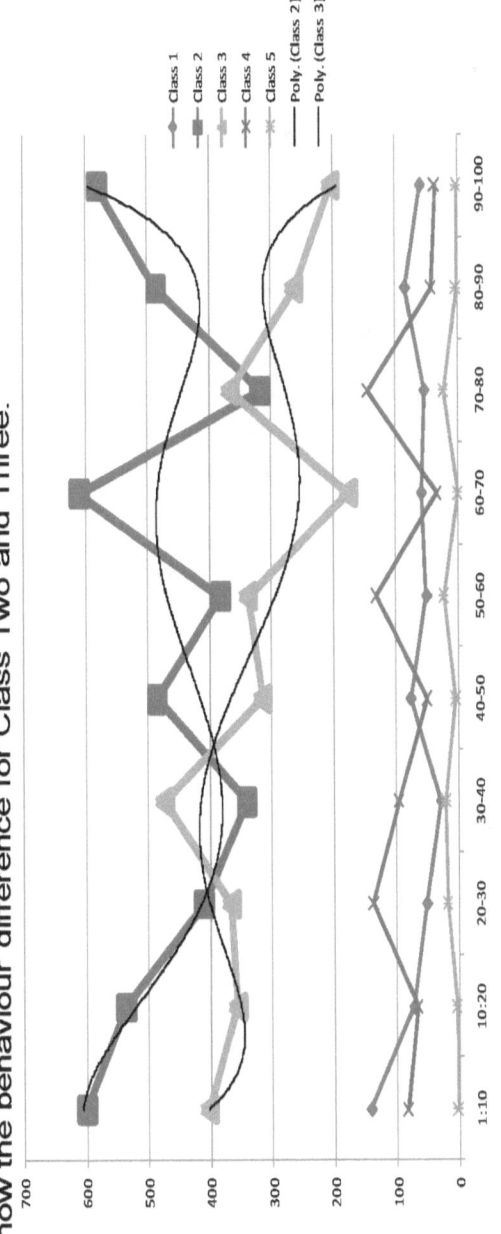

Figure 8. Distribution trends for each prime class. From a statistical observation, the following conjectures may also be made:

4 CLASSIFICATION OF PRIMES

Conjecture 2

There exists no standard primes for $c \geq 7$.

Conjecture 3

The most populous class is that of class 2 primes.

Conjecture 4

There are more left operation primes than there are right operation primes.

The first conjecture is based on the fact that there is a reduction of pristine primes as the class increase from c = 3. In fact, the conjecture is the same statement as the hypothesis that $c \leq 7$.

5 Prime Waves and Event Signatures

The following hypothesis is stated on the basis that the formation of primes is sequential, random, and progressive in pattern.

Hypothesis 2

> Let an F-prime and N-prime represent an event, where an event is defined to be an occurrence with consequent influence and further interaction or relation.

This leads to the concept of mapping a prime referred to as a prime trace or finding an event prime signature. The idea of using an event is based on the fact that events are unique and always affect their environment, even in a small way. Therefore, the suggestion of the hypothesis is that the pattern of the event prime signature may represent the same pattern of an event and how it develops with time into a sequence of events. So in this approach, each prime occurrence represents an event.

Postulate 1

> No two event signatures are the same.

This may be observed from the fact that primes are not predictable through the use of a formula (so far) but exhibit a random nature. Furthermore, the following definitions are made:

Definition 8
a. F-primes represent events that all occur under an original influence.
b. N-primes represent events that all occur under a related influence.

Therefore, the first prime in the sequence represents the first event from which the chain of events begins, and it is called the original event. In a real life, the happening and interpretation of events may be traced in a sequential manner by looking at the original event. In the same way, some events are consequences of a related influence. It is typical to observe such for example in business and markets, where information has direct impact upon business and market events.

Definition 9

> *An event signature is originated from a predetermined prime number that appears as a root in consequent primes.*

The definition effectively summarizes the idea of a prime signature. The research question then is this:

> *Can we use prime roots in the way they occur to represent events and scenarios of events?*

We can probably theorize how the concept of prime roots may be used to create scenario thinking in terms of events. Therefore, a theoretical framework is proposed here, and it may be developed further if it yields other interesting ideas. Primarily:

Definition 10

> *Each event has four variables that define its characteristics, these being result and consequence and impact and influence.*

Every event has some result, and every result has some consequence that we can observe. Similarly, every event will create some impact from its consequence. It is normal in everyday life to consider impact as being the same as consequence. Impact is taken as the transformative power of an event, where this then translates into influence. In analogous terms, this is the same as energy having a certain level of power. Events can be conceptualized as carrying certain social and spiritual energy that creates the desired impact and influence upon people.

THE THEORY OF PRIME NUMBER CLASSIFICATION

Definition 11

Let there be a given prime root event R in p and x the remaining portion of the prime number, then,

a. *R represents the initial prime number*
b. *$p = xR$ or $p = Rx$ is an F-prime if x is odd*
c. *$p = xR$ is an N-prime if x is even*

The major difference here is the R need not be the biggest prime root in a given prime number. Hence, we consider a range of primes satisfying the above condition. For example, we can consider the range {59, 6959}.

Originating Prime	{a = 59, b = 6959}	Total Events for Range
F-primes	359, 593, 599, 1559, 1759, 3259, 3359, 3559, 4159, 4259, 4759, 5903, 5923, 5927, 5939, 5953, 5981, 5983, 6359	19
N-primes	659, 859, 1259, 1459, 2459, 2659, 3659, 5059, 5659, 5987 6659, 6959	12

Table 5. The table shows the N-primes and F-primes, and number of events.

The above primes are then represented in diagrammatic form to derive an event signature.

The following can be observed:

a. **Prime Waves**: Prime numbers that follow each other in succession in the same group define a prime wave, where this can be an N-prime wave or an F-prime wave.
b. **Wave Interruption**: These two prime waves are called the result waves, and a very short wave is called an interruption.
c. **i-Prime Wave**: Prime numbers that follow each other in succession but cross between the group define the i-prime wave, which is an impact wave.

5 PRIME WAVES AND EVENT SIGNATURES

d. **Wave Influence**. The number of prime numbers for a given range divided by the number of crossings between the F-primes and N-primes is called the influence of the event {a,b}.

e. **Connecting Arrow**. Each connecting arrow represents a consequence. The sum of consequences is a result.

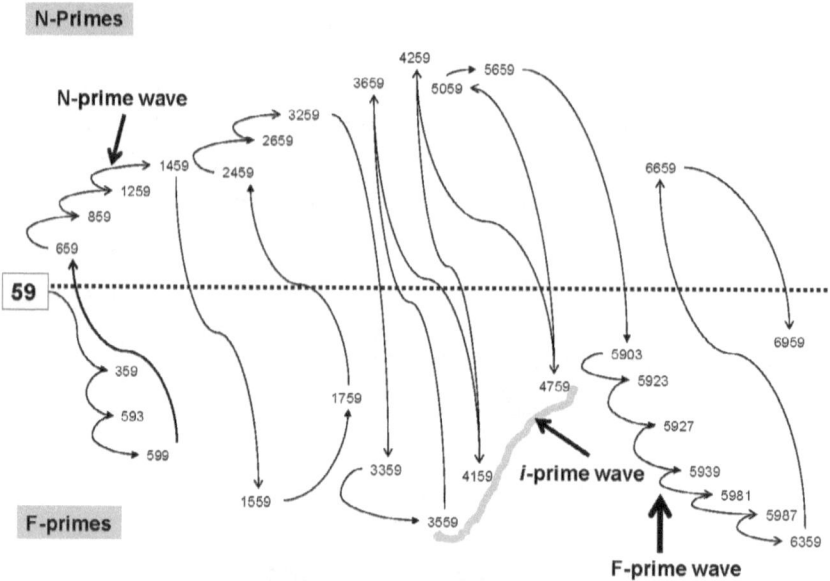

Figure 9. The different type of prime waves.

An event signature can be assigned an interpretation. This is done by looking at the relationship between the different types of waves and how they are occurring. An ideal event signature must have all three types of waves, and an i-prime wave is the most desirable because it implies an occurrence has an impact. For the given range, we can also consider the number of original events, and the events that come as a result of the influence of the other events. For example, there are nineteen original events for this event signature, followed by twelve occurrences that result as an influence from or by the original events.

THE THEORY OF PRIME NUMBER CLASSIFICATION

Prime numbers are random, so this implies that occurrence of prime roots may also be random. As a result, an event signature could probably be used as a diagrammatic approach to trace the development of events by assigning an initial event given a prime. Political events may lend themselves to this type of description or even financial crises management.

A similar approach may be used by using standard primes and pristine primes as shown by the example below for the prime number 409.

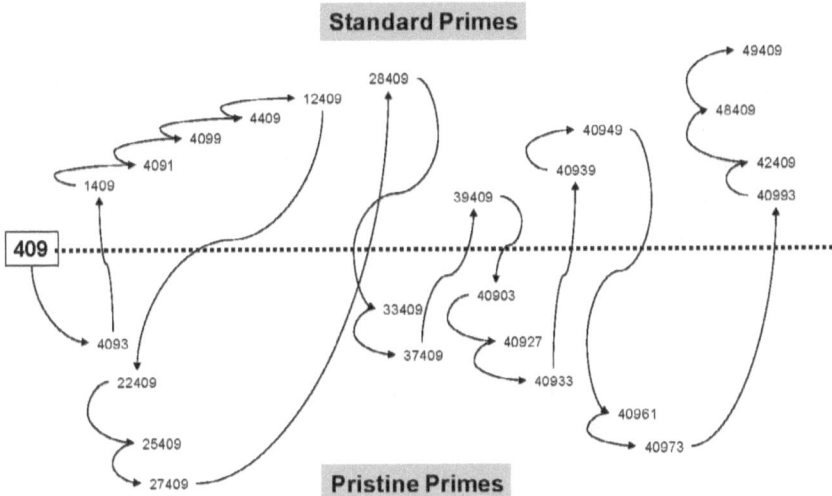

Figure 9b. A similar example with standard and pristine primes.

6 Prime Reactions

The idea of a prime number chain explosion is derived from the fact that a prime is either made from another prime, an odd root, or an even root, and the model seeks to establish some reverse pattern to the process. What the model shows at the center is the initial prime root that is responsible for initiating the reaction. The rules of the reaction are as follows:

> **Rule 1**: The initial prime root can be any prime number.
>
> **Rule 2**: The prime root joins with either an odd root or an even root to form a new prime number.
>
> **Rule 3**: The new prime number becomes a new root to continue the reaction.
>
> **Rule 4**: A reaction exists in a range defined by a begin prime and an end prime.
>
> **Rule 5**: A prime number can be a root once and once only.

These rules form a framework for conceptualizing and studying the prime number chain reaction process. There are variations that one may make in defining a chain reaction depending on what one wants to achieve. The above reaction rules may be defined to form truncatable primes for example.

As a process, the reaction is formalized with this notation:

$N0, N1, N2, \ldots Ne$

where $N0$ is the initial prime, $N1$ the first resulting prime, $N2$ the next until the end prime Ne.

THE THEORY OF PRIME NUMBER CLASSIFICATION

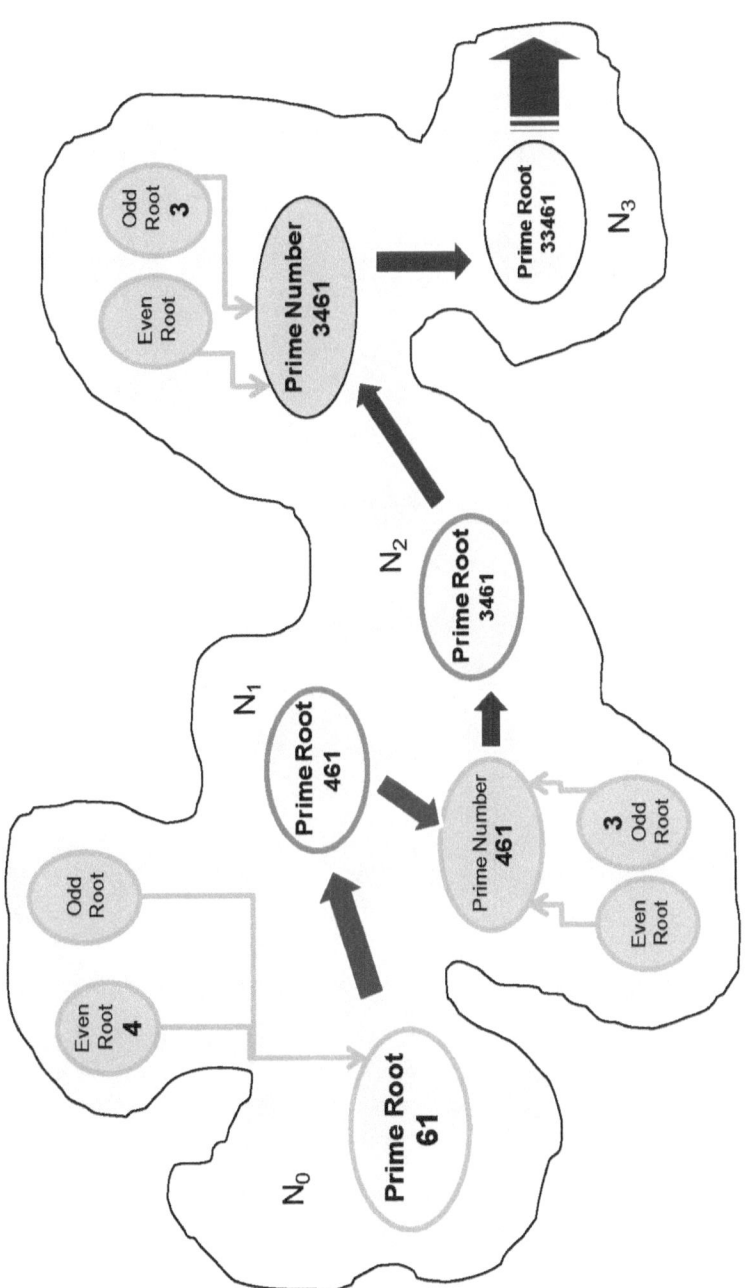

Figure 10. The prime reaction.

Section Two

Positional Classification of Prime Numbers

Overview

Prime numbers occur in a sequential manner, and in such a context, there is no need to consider positional classification. However, by conceptualizing a two-dimensional prime space, then primes can be defined to exist in a given dimension.

Objective:

>To study distributive behavior of prime numbers in a two-dimensional space

Research Question

>*Is it possible to define a concept of a prime space that is two-dimensional as a tool for distribution analysis?*

1 Introduction

Whilst primes are usually studied for their sequential properties in a linear manner, a two-dimensional nonlinear relational approach is applied to develop predictive tools for understanding primes and classifying them. The positional space is defined in order to be able to construct the delta prime space. The definition allows prime behavior to be investigated in terms of the mean patterns and a prime counting technique.

Primes are normally classified according to the type of number patterns that they produce, the effect of interchange of digits in some prime numbers, the reflection of digits within a prime number, and that some group of primes form interesting geometric number patterns. Therefore, positional classification is not a standard approach in understanding prime number behavior, but it can be a useful tool if dimensionalisation is introduced.

2 Defining the Prime Space P_s

It is normal to assume numbers to exist in a one-dimensional space; hence, the common natural geometric presentation of numbers in a real number line. The same argument applies in considering prime numbers; the typical assumption is that they exist in a one-dimensional space. It is not a continuous space as primes are considered random occurring numbers. An alternative approach is to assume that prime numbers exists in a two-dimensional space and then study their behavior in the context of such a space.

> **Axiom 1**
>
> *Prime numbers exist in a two-dimensional space to define the universal set of primes.*

This makes a fundamental difference in the way the primes are presented and allows us to look for other patterns of prime behavior by studying the relationships that arise as a result of the presentation structure. In particular, infinity also assumes a two-dimensional context for the primes rather than a one-dimensional context provided by the sequential approach.

We consider the interval $[m_1, n_1]$ as a prime subspace, where the first parameter is a lower limit and the second parameter an upper limit as an interval consisting of primes from m_1 to n_1. The prime subspace $[m_1, n_1]$ is defined to be a base for other prime subspaces $[m_2, n_2], [m_3, n_3]$ up to $[m_s, n_s]$, where the difference between the lower and upper limit remains constant. Hence the value of s represents the number of subspaces under consideration, where the collection of all subspaces defines some universal prime space.

THE THEORY OF PRIME NUMBER CLASSIFICATION

Definition 1

A prime space consists of elements p, called prime numbers, and of subsets referred to as prime subspaces $[m_x, n_x]$ such that,

a. *A prime space $[m, n]$ is defined to consist of primes from a lower limit m to an upper limit n, where m and n are not necessarily prime.*
b. *A prime space is defined as $P_s = [m_1, n_1, s]$, where $[m_1, n_1]$ defines the base for all other consecutive subspaces $P_x = [m_x, n_x]$ from $1 \leq x \leq s$ such that s gives the span of the prime space.*

The variable n_1 is called the depth of the space provided $m_1 = 1$ as it defines how far down you go for a given column which describes the subspace. The relationship between the depth of the space and the upper limit n defines the span of the prime space. In fact, if d is the depth of the space, then,

$$n = ds = n_1 s \text{ for } m_1 = 1 \quad (1)$$

Similarly,

a. $[m_1, n_1]$ defines the subspace $x = 1$, which is the base of the prime space
b. $[m_2, n_2]$ defines subspace $x = 2$
c. $[m_s, n_s]$ defines the last subspace $x = s$ for the prime space P_s.

Hence, subspaces of P_s must be consecutive for $1 \leq x \leq s$. For example, $[m_1, n_1] = [1, 100]$ is a base for primes in the following:

$$[1, 100], [101, 200], [201, 300], \ldots, [m_s, n\,..]$$

Hence, the interval $[1, 100]$ acts as a base to derive consequent groups, and the depth of the prime space is $d = 100$. Note that the value of d is not a prime number and is normally defined as a power of ten. For example, if $s = 4$, then the prime space is defined as consisting of a group of primes in each of the following subsets:

69

$$[1, 100], [101, 200], [201, 300], [301, 400]$$

That is,

a. the prime space is [m,n] = [1, 400],
b. the prime base is $[m_1, n_1]$ = [1, 100],
c. and the last subspace is $[m_4, n_4]$ = [301, 400].
d. $n = ds$, that is, $n = 100(4) = 400$.

This information is consolidated through the notation that the prime space is defined as,

$$P_s = [1, 100, 4].$$

Prime spaces are therefore different because of their span and the definition of the base that constructs them. Hence, the axiom allows us to create different types of prime spaces depending on what boundary we want to look at.

3 The Default and Standard Prime Space

It is necessary to first establish a positional space of primes in order to develop a theory of delta classification of primes. The positional space merely assumes that every prime has a specific coordinate in the two-dimensional set of primes.

> **Axiom 2**
>
> *In a given prime space, every prime number has a unique position that can be identified as a prime number coordinate.*

Hence, by establishing the two axioms, some basic contextual system of describing the prime numbers is developed. This is an exact location of the prime in the defined space.

> **Definition 2**
>
> *For a given prime space $P_s = [m_1, n_1, s]$, the coordinate of the prime number p in the space is defined as (x, y) where $1 \leq x \leq s$ denotes the subspace and $1 \leq y \leq d$ the sequential position in the subspace.*

For example, in the space $P_s = [1, 100, 4]$, the prime number 353 has the prime coordinate (4, 9), where the first coordinate describes the subspace in which it exists, and the second coordinate describes how far down from the top the prime number is. The definition is said to give the positional classification of primes—each prime is classified according to the subset in which it is found and its position in that subset. Therefore, the structure of sequence in a prime

3 THE DEFAULT AND STANDARD PRIME SPACE

space is downward along a subspace for the prime number, but x describes the specific sequence of subspaces. Consequently, this derives an axiom.

Axiom 3

In positional classification of primes, a subspace represents a positional dimension.

The importance of this lies in the fact that the x variable can be graphed like it is a number instead of just being a number for a set. In other words, we can look for a particular statistical behaviour in the subset x when we treat it as a dimension, and we can graph the outcomes and make a comparison.

The same number 353 has a different coordinate in the space $P_s = [1, 10^5, 30]$. In this case, its coordinate is (1, 71). So coordinates are a function of the defined space. For the sake of convention, it is for this reason that in positional classification, the space $P_s = [1, 10^5, \infty]$ is described as the standard prime space so that any coordinate refers to such a space unless otherwise stated.

Definition 3

Let a prime p exist in a two-dimensional space $P_s = [1, 10^5, \infty]$, where this defines a standard prime space in positional classification.

On the other hand, it may be noted that $P_s = [1, \infty, 1]$ defines the typical prime space that is normally studied, where this is an arrangement of the primes in a sequential linear form. It consists of one column only to define

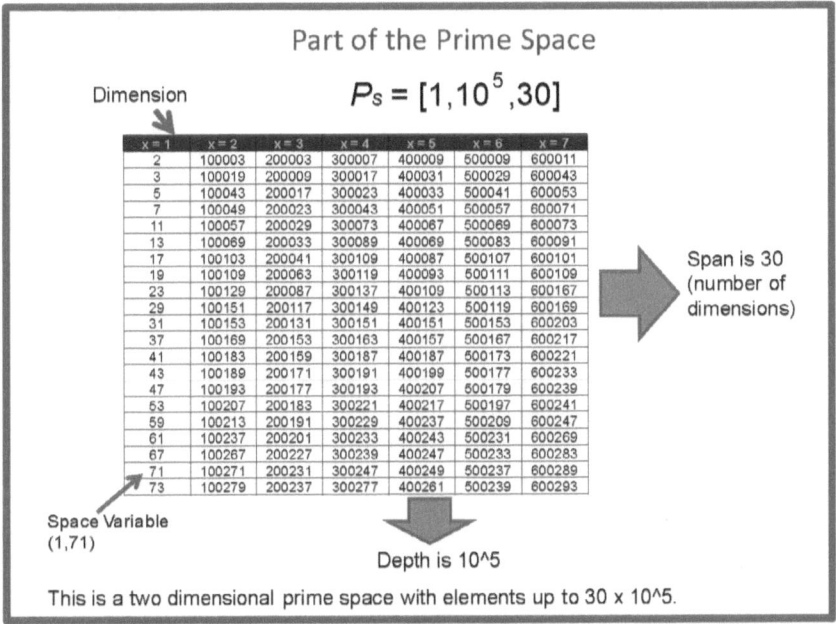

Figure 11. Example of a two-dimensional prime space.

the prime space; hence, it is one dimensional. Hence, the subspace $P_s = [1, \infty, 1]$ defines the whole set—all the elements are in the first subset. This is referred to as the default prime space. The position of the prime 353 is such a space is still (1, 71). An alternative to the default prime space is the definition $P_s = [1, 1, \infty]$. In fact, this is the same the prime space $P_s = [1, \infty, 1]$ in terms of the elements, only that it is along the horizontal rather than the vertical description. That is, in $P_s = [1, 1, \infty]$, each element defines its own subspace. Hence, the coordinate of 353 in this space will be (71, 1).

Similarly, the smallest prime space is defined by P_s = [1, 1, 1] consisting of the prime number element two. On the other hand, $[m_1, n_1]$ = [1, 1] defines the smallest base to describe a prime space and consists of a single prime number. The space of this type consists of one element only, and all such space exists for $y = 1$ since $d = 1$.

Assume a prime space that contains all the primes, where this is the universal prime space defined as $P_s = [1, \infty, \infty]$. Then such a prime space contains all prime numbers in a two-dimensional presentation of infinity instead of one-dimensional structured infinity. This is the difference that axiom 1 makes.

Therefore, it must be assumed for practical reasons that any subspace existing in $P_s = [1, \infty, \infty]$ then $d < \infty$ for the subspace. Also, it is a question of interest whether really $P_s = [1, \infty, \infty]$ exists, or it is merely an abstract notion.

4 Group Behavior of Primes

The aim of dividing primes into specific space and assuming that any given prime has a coordinate is to create a context for comparison. It also provides a platform to derive the delta space of primes. The prime numbers may be studied in their sequential structure, and alternatively, they may be studied in terms of their number group patterns.

Postulate 1

Let there be prime spaces with base $[r, s]$ and $[u, v]$ where $r = u$ and $s \neq v$, then such space will define the same statistical behavioral characteristics.

This postulate provides a basis for relating a group of primes or for finding a result in a given group and allows that assumption that it will be true for another prime group.

In order to compare two prime spaces, then $r = u$ implies that the prime spaces have the same initial point. The upper limits of such space may differ, but in order to compare their characteristics according to the postulate, then their base must differ, hence the condition $s \neq v$. That is, the prime spaces will differ in their depth. Consequently, the assumption $r = u$ is a standard approach, variations to the postulate may be considered for $r \neq u$. But for a start, this is the context on which the postulate is based.

If we find a particular statistical function in [r, s], we expect to find a similar function in [u, v], provided both are subsets of $P_s = [1, \infty, \infty]$. Therefore conversely, if two prime spaces have the same characteristic, then it implies that they are a subset of a larger space in which that characteristic is true.

4 GROUP BEHAVIOR OF PRIMES

The implication is that a statistical finding for a given subspace should reflect a characteristic of the universal prime space, which implies that one need only investigate at a finite level to postulate a similar behavior.

An example is given to illustrate the assumptions of the postulate. Assume the prime space $[m, n] = [1, 4000]$. The requirement is to count the number of primes per hundred in all the respective subgroups up to 4000. Dividing by hundred gives us forty such subspaces, that is $s = 40$. Hence, the prime space is denoted as $P_s = [1, 100, 40]$. A count may be carried out where the result is summarized as follows:

Dimension	$[m,n]$ for $d = 10^2$	Count : $d = 10^2$
x = 1	[1, 100]	25
x = 2	[101, 200]	21
x = 3	[201, 300]	16
x = 4	[301, 400]	16
	continues to	
x = 37	[3601, 3700]	13
x = 38	[3701, 3800]	12
x = 39	[3801, 3900]	11
x = 40	[3901, 4000]	11

Table 6. Counting primes in a two dimensional space.

Since $1 \leq x \leq s$ then $x = 1$ denotes the subspace [1, 100], $x = 2$ denotes [101, 200] and so on till $x = 40$ for [3901, 4000]. A trend is observable when a graph is plotted of the count for each group, as shown by the logarithmic trend line on the results. It is a decreasing statistical function.

Now consider $P_s = [1, 10^5, 30]$. Hence, $[1, 10^5]$ defines $x = 1$, similarly $[1, 2 \times 10^5]$ defines $x = 2$, and so on. This derives a table as shown with a downward trend as x increases. In other words, it is also a decreasing statistical function. The values of the functions differ for each x, but the statistical characteristic of decreasing is true for both of them. This is the essence of the postulate.

Dimension	Count: $d = 10^5$
x = 1	9592
x = 2	8392
x = 3	8013
x = 4	7863
... continues	
x = 27	6717
x = 28	6747
x = 29	6707
x = 30	6676

Table 7. Counting for a given prime depth.

Remark 2

A decreasing count function in the prime space corresponds to the fact that as one goes toward infinity, the gaps between prime numbers deceases. Hence, if one has a fixed subspace, then the count of primes should decrease as *x* increases.

Whilst the resolution of the behavior may be clear for a small interval as indicated by the previous graph, larger intervals of [m, n] define a smoother curve. In each of the two cases presented for different primes spaces, a clear discernable downward trend is derived, the number of primes per 100000 decreases. The same is true if one considers it in the context of number of primes per hundred. If a statistical behavior is true for a space [r, s], then it can be assumed to be true for another space [u, v] as long as they exist in the same universe.

Remark 3

The application of this is that one can prove a given behavior at smaller space with the implication that is true for a larger space. Since primes continue to infinity, this observation is useful in that it allows one to make an observation that is true at manageable values of primes. In other words, it is similar to an inductive process, if it can be proven true for case n and for case $(n + 1)$, then it is true for all cases.

Secondly, the postulate illustrates the application of axiom 1 because it allows x in the range $1 \leq x \leq s$ to be treated like it is a single variable even though it is actually representing a group. The usefulness of this is that we may group primes and develop equations in x that are used to describe group behavior of primes. So it must be noted that throughout, x is a variable that stands for group behavior. So for example, whereas in $Ln(x)$, x is actually a prime number in the prime number theory, here, $Ln(x)$ means the log of the group number x in the sequence of the defined prime space.

5 The Prime Number Count Equation

The above postulate can be applied to develop a prime number count equation. As distribution theory of primes usually indicates, the number of primes per hundred thousand decreases very slowly. This is evident from the downward trend of the figure 12. The graph shows the count for the prime spaces:

- $[m, n] = [1, 3000000]$

- $[m, n] = [7000001, 7100000]$

- $[m, n] = [10000001, 10100000]$

- $[m, n] = [30000001, 30100000]$

- $[m, n] = [50000001, 50100000]$

- $[m, n] = [80000001, 80100000]$

- $[m, n] = [999000001, 999100000]$

5 THE PRIME NUMBER COUNT EQUATION

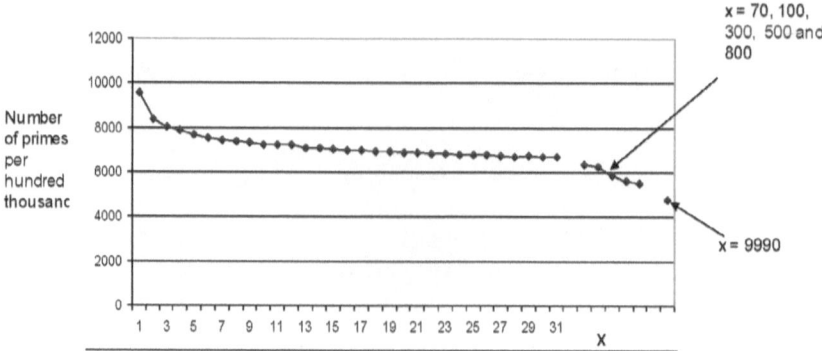

Figure 12. Using dimension for prime count.

Similarly, it can be observed that the drop in count of the number of primes for the space [m, n] = [999000 001, 999100000] is not significantly large. The fact that primes decrease very slowly allows the development of mathematical formulae to describe them in spite of their random behavior.

Consider the prime space $[1, 10^5, 30]$ = [1, 3 000 000], and a count per hundred thousand is done on the prime space, then the logarithmic trend line derives the equation:

$$y = -689.94 Ln(x) + 8947.5$$

where x is the group number in the prime space sequence. From $x = 2$ to $x = 11$, the formula estimates the count of primes per hundred thousand as being slightly above the actual number and then follows a very tight close relation onward that reflects a satisfying degree of accuracy. However, as x increases, the variation begins to grow, and the formula is more inaccurate. This is due to the influence of the constant multiplier 689.94 on $Ln(x)$ and intercept value 8947.5, which may not take into account the fact that prime numbers behave in a random fashion. The technique applied then is that of defining a statistical outlier for some of the values of x. The count at the beginning is then assumed to represent statistical outliers, and they are removed one by one in order to find the average behaviour the multiplier on $Ln(x)$ and the intercept. The prime counts that are systematically removed and taken as statistical outliers are 9592, 8392, 8013, 7863, 7678, 7560, 7445, 7408, 7323, 7224, 7216, 7224, 7083, 7105. The derived figures are as shown in Table 8.

Hence, we can now replace the constants 689.94 and 8947.5 with estimates that depend on the value of x. The focus of the formula is to try and achieve more accuracy as x increases, hence the modification by making the formula more sensitive to changes in x.

Prime Count	Multiplier on Ln(x)	Intercept
9592	689.93	8948
8392	580.14	8640
8013	553.71	8564
7863	543.37	8534
7678	527.3	8487
7560	518.76	8461
7445	512.11	8441
7408	511.9	8440
7323	502.73	8412
7224	497.75	8396
7216	506.94	8425
7224	504.38	8417
7083	477.62	8332
7105	489.66	8371
7029	474.32	8321
Trend Line	$y = 647.88x^{-0.1143}$	$y = 8830.5x^{-0.0219}$

Table 8. Deriving a prime count equation.

The count formula per hundred thousand then becomes,

$$y = 647.88x^{-0.1143} Ln(x) + 8830.5x^{-0.0219} \quad (3).$$

Comparisons and accuracy are demonstrated by calculating values for x = 70, 100, 300, 500, 800, and 9990. These values are multiplied by 10^5 to get the actual number that they represent for the prime space. For example, x = 70 stands for $7000000 = 70 \times 10^5$, and this implies the prime space is defined as $[m, n] = [7000001, 7100000]$.

5 THE PRIME NUMBER COUNT EQUATION

x =	70	100	300	500	800	9990
Unmodified formula: $y = -689.94 Ln(x) + 8947.5$	6016	5770	5012	4660	4336	2594
Modified Formula: $y = 647.88x^{-0.1143} Ln(x) + 8830.5x^{-0.0219}$	6352	6221	5868	5728	5611	5135
Actual prime count per 100 000	6367	6241	5860	5592	5496	4760
Difference from Modified Formula	15	20	-8	-136	-115	-375
Difference from Unmodified Formula	351	471	848	932	1160	2166

Table 9. Comparing accuracy of equations.

Within the context of randomness of primes, the modified formula appears to make a reasonable estimate of the primes per hundred thousand depending on the level of error that ones to achieve.

6 The Mean of Primes

In finding mean of a collection of prime numbers, one consideration is that primes may be added in a sequential manner and their mean taken. This gives a diverging figure as the prime numbers increase toward infinity. In the two-dimensional space, we can consider subsets, and since each subset forms a variable in the prime space, this allows comparison for the behavior of the mean.

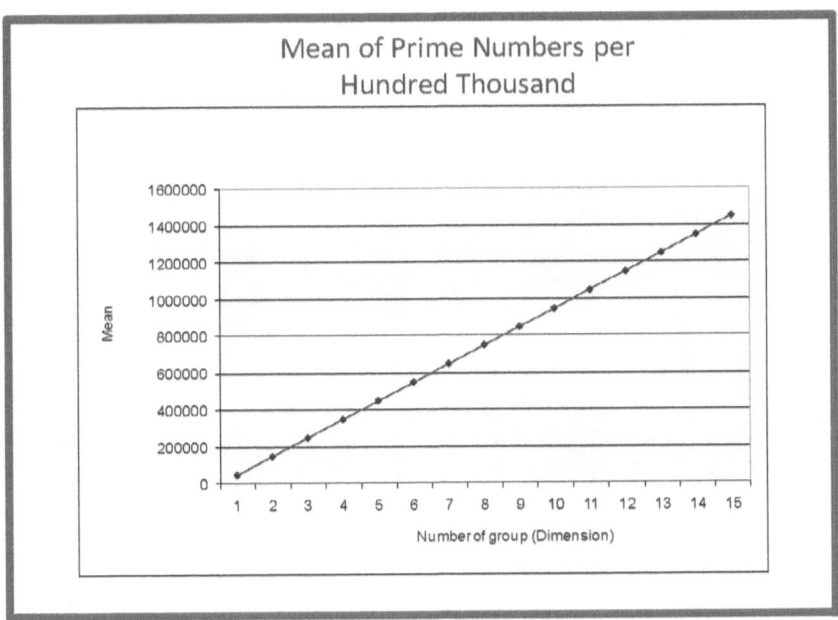

Figure 13. The means of a group of primes lie along a straight line.

Consider the function $y = f(x)$, where $x \geq 1$ is the variable describing the set, and y is the mean of all the primes in x, then we derive the following graphical points:

6 THE MEAN OF PRIMES

$(1, 47372), (2, 149572), (3, 249583), (4, 349662), (5, 450031),$

$(6, 549946), (7, 649894), (8, 749916), (9, 849880), 10, 949760),$

$(11, 1049799), (12, 1150007), (13, 1250119), (14, 1350213),$

$and (15, 1450169)$

These derive a best line fit of $y = 100094x - 51026$, where $x \geq 1$. In other words, between the subspaces, there is a linear relationship of the means of the primes per set. This is an interesting property considering the fact that primes are random in nature. Even if a different space is defined, the same observation will still stand, confirming the observation made by in the first postulate.

This shows a stunning regularity of the mean for any x to the extent that it is almost a perfect straight line. This confirms the observation made that prime numbers, though random, can also be described in a precise manner in terms of their behavior.

7 Average Gap in a Sample Subspace

It is generally assumed from the prime number theorem that the average magnitude g_n of a gap between two primes p_1 and p_2 is $Ln(p_1)$). We start by making a definition of another type of subspace in a given prime space.

> **Definition 4**
>
> Consider a prime space $P_s = [m, n, s]$, then a sample subspace is defined as $P_u = [u, d, s_1]$ where u is an initial prime and d the number of primes sampled per prime space prime and $s_1 < s$.

For example, consider the prime space [1, 105, 15]. Then from that, we choose a subspace given by $x = 2$ and from this, select a sample of one hundred primes beginning with 100019. The one hundred defines the depth d of the sample from a given initial prime number. Hence, we define this sample prime space as $P_u = [100019, 100, 1]$ since we are remaining in the same subspace. The last prime number in this sample will be 101197. If we had defined it as $P_u = [100019, 100, 2]$, then it would span over two subspaces, $x = 2$ and $x = 3$, so $x = 2$ would be an initial subspace, and the last prime number in it would be 201287. The notation just describes the sample prime subspace, but it can be chosen at random. Since the largest space of the prime space is given by s, it implies that the choice of s_1 cannot be greater than s.

7 AVERAGE GAP IN A SAMPLE SUBSPACE

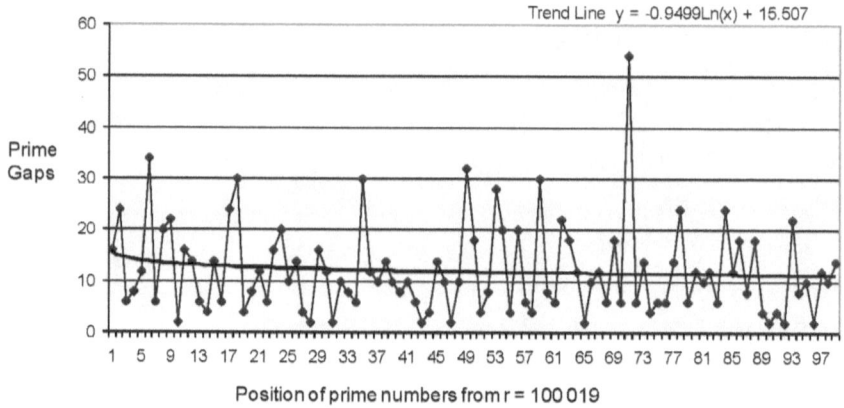

Figure 14. Prime subspace distribution of gaps.

Choosing a sample prime subspace can be used to study the behavior of prime numbers at a particular portion of a prime space, and it is also emphasized that the notion is that such sample subspaces should be randomly chosen.

Conjecture 1

> For a given prime space $P_s = [m, n, s]$, the average gap of the primes in the sample subspace $P_u = [u, d, s_1]$ will be given by $y = aLn(x) + b$ where a and b are some constants.

Now consider picking up a prime subspace defined as $P_u = [u, d, s_1]$, where we choose such at random in some prime space at $x = 2, 70, 300, 800$ and 9990 from $[m, 105, 1]$. The subspace $[u, 100, 1]$ is picked up from within the prime space $[m, 105, 1]$. A trend line and formula is then derived for each such subspace, where this gives the average behavior of the prime gaps. The value from the formula is found at $x = 50$, halfway one hundred of the prime subspace. This is compared to the actual computed average prime gap for the sample space and $Ln(q)$, where q is the average of $(u + v)$, where u is the first prime number, and v is the last prime number. Since primes are random, some variation is expected.

All three forms of approach approximate each other, that is, the actual calculated average, the use of the formula, and the calculation of $Ln(q)$. The formulas are derived from the trend lines for the subspace where $d = 100$ in each case.

THE THEORY OF PRIME NUMBER CLASSIFICATION

x =	Actual Average	Value and Formula for x = 50 in the range 1 to 100	$Ln(q)$	u	v
2	11.91563	11.79097 y = −0.9499Ln(x) + 15.507	11.51899	100019	101197
70	15.7041	16.93198 y = 0.5777Ln(x) + 14.672	19.6834	7000009	700001663
300	17.08479	17.50865 y = 2.3216Ln(x) + 8.4265	17.21674	30000023	30001669
800	18.18886	20.51953 y = 0.7624Ln(x) + 17.537	18.19755	80000027	80002033
9990	21.0042	24.00059 y = 1.6346Ln(x) + 17.606	20.72227	999000029	999002341

Table 10. Showing the mean gap value calculations.

SECTION THREE

DELTA CLASSIFICATION OF PRIME NUMBERS

Overview

A particular approach is using gaps to define a prime family in a given prime space interval P_f and then classify the primes according to their gaps. Hence, it is possible to establish seven types of gap analysis techniques by developing a concept of the "gap in the gap" of primes using prime families and to demonstrate that prime gaps have a positive, negative, and steady rate of gap change referred to as gap acceleration. Similarly, using comparative gap analysis, the average gap between twin primes is increasing at about nineteen times faster than the average gap between the consecutive primes.

Objective:

To study distributive behavior of prime numbers with respect to a prime space defined by the natural prime gap.

Research Question

Is it possible to do a descriptive analysis of primes in terms of gaps influencing the prime distribution rather than primes influencing the gap distribution?

1 Introduction

Therefore, one may consider categorizing the primes into groups that are defined as prime families within a given subspace. The concept of a family is based on prime gaps, where each gap defines a family. Consequently, this offers a different way of classifying primes and is referred to as the Delta Classification System where the word "delta" denotes difference.

2 The Prime Family

A prime space is defined as $P_s = [m_1, n_1, s]$, where $[m_1, n_1]$ defines the base for all other consecutive subspaces $P_x = [m_x, n_x]$ where $1 \leq x \leq s$ such that s defines the span of the prime space. The span is the number of subspaces constituting the prime space. For example, a prime base of $[1, 200]$ means we subdivide primes into groups where the difference between the upper bound and lower bound per group is two hundred. If we have a span of 10, meaning ten subsets, hence the prime space $P_s = [m_1, n_1, s]$ starts from 1 to 2000 and is denoted $P_s = [1, 200, 10]$. Therefore, $x = 1$ is the subset of primes from 1 to 200, $x = 2$ the subset of primes from 201 to 300, and this continues to 2000.

Generally, the nth prime gap g_n is defined as the difference between the $(n+1)^{th}$ and the nth prime number, that is, $g_n = p_{n+1} - p_n$. For the purpose of distinguishing concepts and clarity, this is also referred to as the natural gap of the set of prime numbers. This definition assumes the natural consecutive arrangement of primes such that it is a continuum, that is, we are looking at one big prime space. However, the prime space P_s consists of subspaces such that we may consider the position of a gap with respect to a given subset only, that is, the nth prime gap for $x = 4$, for example. For the delta space that will be defined in a moment, there is no interest in establishing the nth prime gap because in such a space, the gap is a constant that defines a prime family. Therefore, the prime gap is only considered in the context of magnitude rather than as the number of composites before the next prime.

> **Definition 1**
>
> The delta prime gap f_g is the difference between a prime p_n and the previous prime p_{n-1} in a given group of primes such that f_g

describes a family (p_x, f_g), where $1 \leq x \leq s$, $f_g = p_n - p_{n-1}$, and $f_g = 2k$, for $1 \leq k \leq \infty$.

In the so-called delta prime space, f_g is even and constant because of rearrangement of the primes, whereas g_n is a random occuring variable, and g_n concerns the next prime while f_g concerns the previous prime, hence the use of the word *delta gap* in the definition. So there is an ideological difference even though the arithmetic definition is similar. The variable k counts the number of families in a given prime subspace.

Definition 2

Let $\Delta = (\Delta_2, \Delta_1)$ be a transformation function that converts the given prime space P_s to P_f consisting of families, such that $(\ddot{A}_2, \Delta_1)(Ps) = \Delta_2(p_n, g_n) = (p_x, f_g)$, where $(p_x, f_g) \in P_f$ defines a family, and x denotes the subspace in which the prime is found.

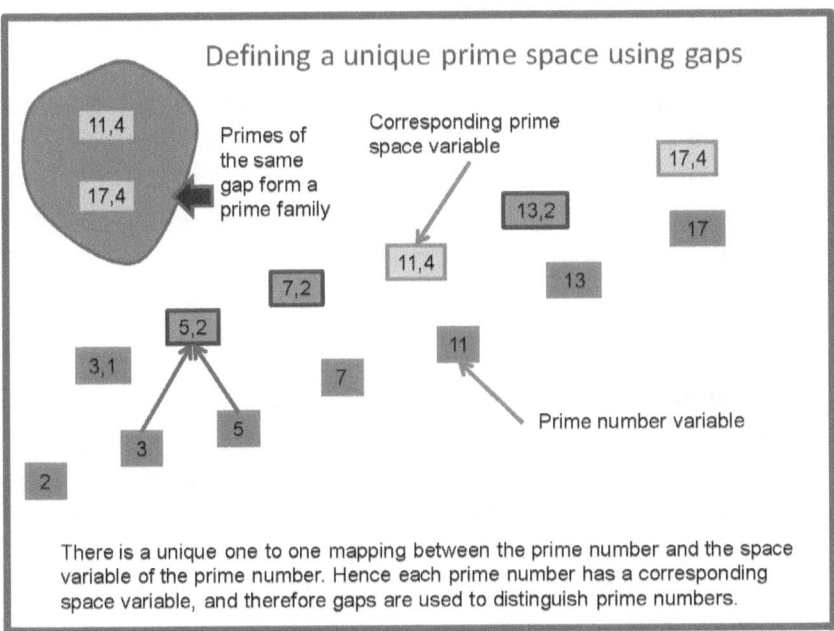

Figure 15. Relating primes in terms of gaps.

Therefore, $\Delta P_s = P_f$; hence, P_f is called a delta space. The primes are no more in their original order, and therefore, the concept of the nth gap is not applicable unless you refer to the nth gap in a given subspace in P_f. The above are best illustrated by an example. Consider prime numbers less than twenty, then,

$$(p_n, g_n) = \Delta_1(2, 3, 5, 7, 11, 13, 17, 19)$$

That is,

$$(p_n, g_n) = \{(2, 0), (3, 1), (5, 2), (7, 2), (11, 4), (13, 2), (17, 4), (19, 2)\}.$$

To understand the rationale, consider for example the difference between 5 and 3 is two; hence, we can either have the following:

a. $(p_n, g_n) = (3, 2)$ meaning in general that the nth prime to the next prime, there is a gap of two, that is, $(p_n + g_n) = p_{n+1}$. This is the conventional approach to gap theory. Alternatively, we can say,
b. $(p_n, f_g) = (5, 2)$ meaning in general that the (n-1) prime has a gap of two to the previous prime. That is, $(p_n - g_n) = p_{n-1}$.

The notation of the definition settles on the second format for conceptual reasons, and the same value of $g_n = f_g$ applies in both situations, the difference lies in the choice of p_n. Therefore, the delta function Δ_1 creates a relationship (5, 2) according to g_n, and this applies to any prime element. The delta function creates a set where the primes have a one-to-one relation through the prime gap of the previous prime number.

Remark 1

Primarily, the delta function creates a two-dimensional perspective of the prime number by making the gap an attribute of the prime. This unique relationship is established between all the prime numbers in terms of a difference to the previous prime, and consequently, it gives a different presentation format to the gap theory that allows us to establish certain observations.

THE THEORY OF PRIME NUMBER CLASSIFICATION

The second delta operator now orders the primes according to the difference; therefore, the difference is constant for a given subgroup called a prime family. The prime space is said to be grouped in terms of their delta families. Therefore, we have,

$$\Delta_2(p_n, g_n) = (p_x, f_g)$$

That is, Δ_2 now sorts the primes in terms of gaps and orders them according to their size. Therefore, the primary sort is not according to the size of the prime number, and this happens within the context of a group, hence the notation p_x. Note that k counts the number of delta families, with the exception of (3, 1) in the first subspace. So $k = 0$ is the first delta family, and $k = 1$ is the second delta family. Therefore,

$$\Delta_2\{(2,0),(3,1),(5,2),(7,2),(11,4),(13,2),(17,4),(19,2)\}$$
$$= \{(2,0),(3,1),(5,2),(7,2),(13,2),(19,2),(11,4),(17,4)\}$$

The primes are ordered according to the gap 0, 1, 2, and 4. To derive a family, we then set for each case $f_g = 2k$. If $x = 1$ for the primes less than twenty, then the families are,

a. $(p_x, f_g) = (p_1, 2) = \{(5,2),(7,2),(13,2),(19,2)\}$ are in the same family since $d = 2$. In this case, $k = 1$ and $d = 2k$;
b. $(p_x, f_g) = (p_1, 4) = \{(11,4),(17,4)\}$ are in the same family since $d = 4$. In this case, $k = 2$, that is (p_x, d) and $d = 2k$.

Therefore, $(p_1, 2)$ and $(p_1, 4)$ are elements of the prime space P_f, while (2, 3, 5, 7, 11, 13, 17, 19) are elements of the prime space P_s.

3 The Discrete Prime Space

Since a prime space consists of subspaces $1 \leq x \leq G$, it implies there are two ways of treating the concept of a prime space, that is,

a. the prime space consists of a discrete entities, which are subspaces with an initial element and an end element; and
b. the prime space is a continuous entity, the initial element of a given prime subspace is always connected to the last element of the previous prime space.

The second option describes the common approach in treating primes and results in a linear approach. In this case, the primes are not studied as a discrete group but as a continuum with an initial prime element two to infinity. For example, you talk of the nth gap in the whole prime space or the behavior of twin primes in the whole space. However, by assuming that the families are discontinuous, this allows the possibility of treating a family as a variable and to study properties of the primes in the context of subspaces. Consequently, one may then generalize on the basis of observations made in given subspaces to the complete prime space.

> **Remark 2**
>
> The ability to make an observation and conclusion on one subspace and to generalize the result for the nth subspace provides a tool for studying primes at a finite space and provides a possibility to make conclusive assumptions for the case at infinity.

THE THEORY OF PRIME NUMBER CLASSIFICATION

The disadvantage of the discrete approach is that it introduces error. For example, in a prime space $P = [1, 10^5, 10]$ then for $x = 9$, the last element is 899981 and the first element in $x = 10$ is 900001. The difference between these two consecutive primes is twenty, but we lose the information $\Delta_1(900001) = (900001, 20)$ because they are in different sets, and a discrete space is assumed. That is, this becomes $\Delta_1(900001) = (900001, 0)$, implying the difference with itself as the first element of the subset is zero. However, if the span of a space is reasonably large, this minimizes the error. In this case, we are losing one piece of information per 10^5. For any set P_f, the initial element is always $(p, 0)$, where p is the first prime number in a given subspace x.

Remark 3

Hence, the consideration is that the impact of losing some information as a result of the discrete approach does not significantly and statistically decrease the interpretation and result of any observations made with regard to the primes as a family.

4 Shadow Family

The fact that $\Delta_2(p_n, g_n) = (p_x, d)$ allows us to trace back a preceding prime number from a given family. Once a given family is defined, one can derive the other family, called the shadow family, which represents the predecessors. For example, the predecessor of (13, 2) is (13-2, 4) = (11, 4). In a predecessor, this gap variable is called a shadow gap, that is, in the prime family defined by two, at the point (13, 2), four is a shadow gap of two. Therefore, the elements (5, 2), (7, 2), (13, 2), (19, 2) have the shadow elements (3, 1), (5, 2), (11, 4), (17, 4). It is also evident that in a given family, the gap between the shadow family and its predecessor is also increasing. For example, consider the family of primes $(p, 2)$. Then (139, 2) has the shadow prime (137, 10). It has a gap of ten from its predecessor. As one gets to infinity, the shadow gap increases in an arbitrary fashion, though no particular study has been made as to how it varies. For example,

Family	Shadow family	Shadow Gap
(300931, 2)	(300929, 36)	36
(401311, 2)	(401309, 22)	22
(601189, 2)	(601187, 40)	40

Table 11. Shadow gaps of primes.

5 Some Applications

Several applications may be derived from the concept of a discrete prime space. For example, there is a general wide interest in the gap of two in prime numbers because of the twin-prime conjecture. The gap between primes per hundred thousand can also be graphed, and we derive comparative patterns of prime number behavior. This leads to other observations that may be formulated with respect to prime gaps.

5.1 Twin-Prime Conjecture in the Prime Space

When using the operator Δ_1, the prime space will have the gaps in an unordered random fashion as expected. However, the operator Δ_2 orders the primes in terms of gaps as the primary sorting order then in terms of size of the primes as the secondary sorting order.

We can then observe two important conditions for the P_f space, that is,

Condition 1

If $\Delta = (\Delta_2, \Delta_1)$, and $P_s = P_f$, where $\Delta_2(p_n, g_n) = (p_x, f_g)$, and $d = 2k$, for $0 \leq k \leq \infty$, then P_f defines a discrete prime space, where the first element will always be $(p, 0)$.

Condition 2

Let the sorting operation by Δ_2 operator be ascending, then it implies the second row in the P_f space will always be $(p_x, 2)$ given by $k = 1$.

The second condition describes the assumption of the twin-prime conjecture, that is, that there will always be a twin prime even at infinity. Hence, after $(p, 0)$ for a given prime subspace at $x = \infty$, the next element in P_f is $(p_x, 2)$. This gives one the impression that even if we continue to infinity, in each subspace, the smallest gap will be two if we ignore the first element of the subspace. That is, in terms of the prime space theory, the conjecture can be restated as follows.

Conjecture 3

In the prime space $P_f = [1, \infty, \infty]$, then $(p_x, 2) \in P_f$.

It must be noted that the space is a two-dimensional structure, so the depth is at infinity, and the same applies for the span. The assumption is that primes are endless at infinity. This makes it an extremely large space and implies that in every subspace of P_f, we have the element $(p_x, 2)$.

5.2 THE PRIME COUNT SPACE GAP

Generally, the prime gap is the difference between two successive prime numbers as had been discussed. The advantage of defining a prime space is that we can now define another type of gap in terms of a horizontal analysis.

Definition 3

For a given prime space, a prime space gap is defined as the difference between two counts of successive prime subspaces, that is, $c_x = P_x - P_{x+1}$ where $1 \leq x \leq s$.

THE THEORY OF PRIME NUMBER CLASSIFICATION

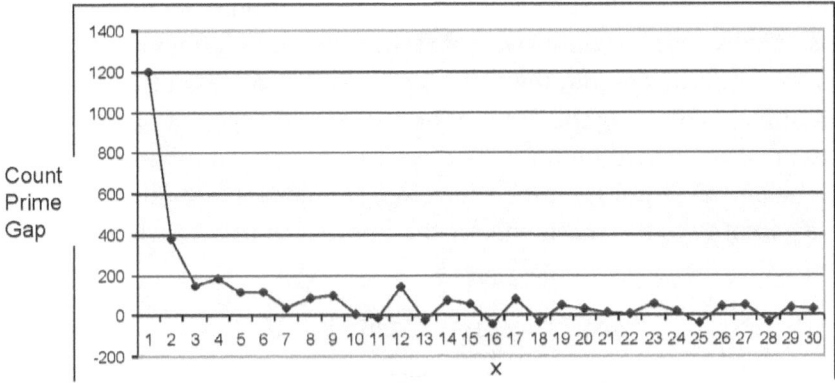

Figure 16. The gap between subspaces.

The difference of a space gap fluctuates around zero; it can be either positive or negative as shown for the prime space $[1, 10^5, 30]$ on graph 14. It may be observed that as x increases, then the prime space gap becomes smaller.

5.3 THE PRIME COUNT FUNCTION $\pi(x, d)$

The fact that primes can be studied with a vertical and horizontal analysis also provides some insights in the behavior of gaps in terms of the family concept in primes. The prime count function is defined as $\pi(x)$, where x is the number of primes below a given prime x. Since the consideration is to apply the same approach to the prime space concept, the change is to define the same function but with a family notation included.

Definition 5

Let $\pi(x, d)$ describe the number of primes in a given family, where $1 \leq x \leq s$, and $d = 2k$, where $0 \leq k \leq \infty$.

In this case, the count is not defined to be less than a given prime, but rather, it is the number of primes in a designated family. For example, $\pi(6, 4) = 775$ means that in the subspace $x = 6$ for the prime gap 4, then there are 775 primes. Note that since this is a discrete prime space, there will always be the possibility of an error of one in the count since we have one element that is always $(p_x, 0)$.

5 SOME APPLICATIONS

The prime count can also be expressed as an approximation equation for gaps. In positional analysis, it was used to find the number of primes in a given family. The same applies here, but the focus is primes with a particular gap. Consider, for example, in deriving the number of primes with a gap of two per hundred thousand. From figure 17, it can be observed that the number of such primes is on the decrease per hundred thousand. On plotting the number of primes with a gap of two, the trend line for the number of primes with a gap of two comes to,

$$\pi(x, 2) = -147.84 Ln(x) + 1060.8$$

Remove	Constant on $Ln(x)$	Intercept Value
—	147.84	1060.8
1224	113.89	971.11
936	104.51	945.63
834	103.78	943.59
810	101.17	936.24
761	104.67	946.25
766	102.36	939.58
730	106.07	950.42
705	116.15	980.15
706	124.93	1006.3

Table 12. Deriving a count equation for gap 2.

The constant on $Ln(x)$ as well as 1060.8 does not provide sufficient sensitivity to the random behavior of the primes, and so a better estimate is made by removing each value from the beginning to obtain an average trend line. This is given by table 12.

From the iteration shown in the table, trend lines are used to derive better estimates that vary with x. The trend line from the points of the constant on Ln(x) is $-125.98x^{-0.0792}$ while the trend line on the intercept value is $1004.9x^{-0.0252}$. The formula derived now becomes,

$$\pi(x, 2) = -125.98x^{-0.0792} Ln(x) + 1004.9x^{-0.0252}$$

THE THEORY OF PRIME NUMBER CLASSIFICATION

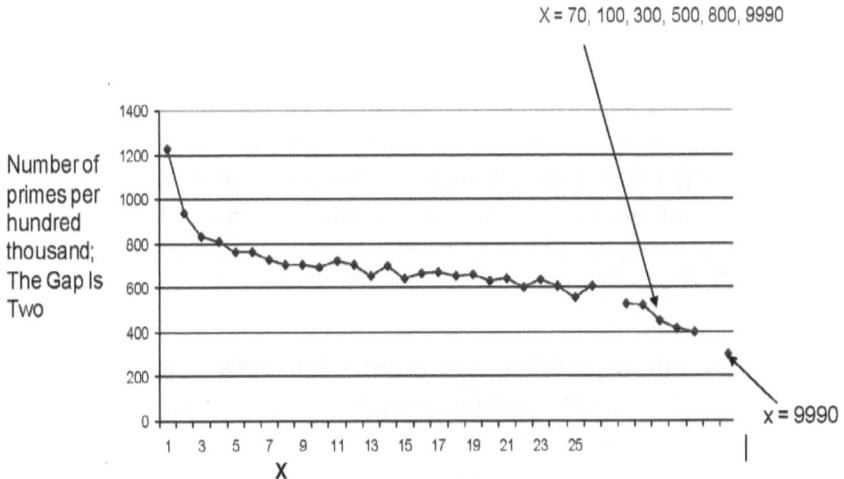

Figure 17. Graph of primes where the gap is two.

This is an estimate of the number of primes in a family defined by the variable x per hundred thousand that have a gap of two. This is tested for values of 7 million ($x = 70$), 10 million ($x = 100$), 30 million ($x = 300$) , 50 million ($x = 500$), 80 million (800), and 999 million ($x = 9990$). It will be noticed that there is a dramatically improved estimate on the formula for the count of the number of primes that have a gap of two compared to the previous formula.

$x =$	70	100	300	500	800	9990
Actual prime count	524	522	454	415	401	299
Unmodified formula	433	380	218	142	73	-301
Difference from actual	91	142	236	273	328	600
Modified formula	521	492	413	381	353	237
Difference from actual	3	30	41	34	48	62
Error per 10^5	0.6%	6%	9%	8%	12%	21%

Table 13. Comparison and accuracy of count equations.

Note the negative value for $x = 9990$ in the unmodified formula. The implication of the negative value is that there is no expectation of finding twin primes there. This, of course, is replaced by the better estimate in the modified formula.

5 SOME APPLICATIONS

5.4 GROUP FREQUENCY DISTRIBUTION PATTERNS

The concept of a prime space is to look at behavior patterns of primes in terms of families in a given subspace and in a two-dimensional perspective rather than a linear approach. The families being discrete can be compared as variables in the prime space. That is, for example $x = 4$ defines a subspace in $[1, 10^5, 30]$, and we can determine the distribution of the families in this subspace. If we consider that only, this becomes a one-dimensional analysis as shown in the figure 18.

Now we can compare families across each subspace, and this gives us a two-dimensional perspective of analyzing prime number properties. This is the major advantage of the two-dimensional prime space concept and is similarly expressed in the positional classification of the primes. The other point of note is the fact that what is true in a given space, a similar statistical behavior may also be described in a different prime space. Therefore, an observation made at a smaller prime space is true also at a larger prime space, implying that one may hypothesize on a smaller scale. The theorems below express group behavior of prime number patterns.

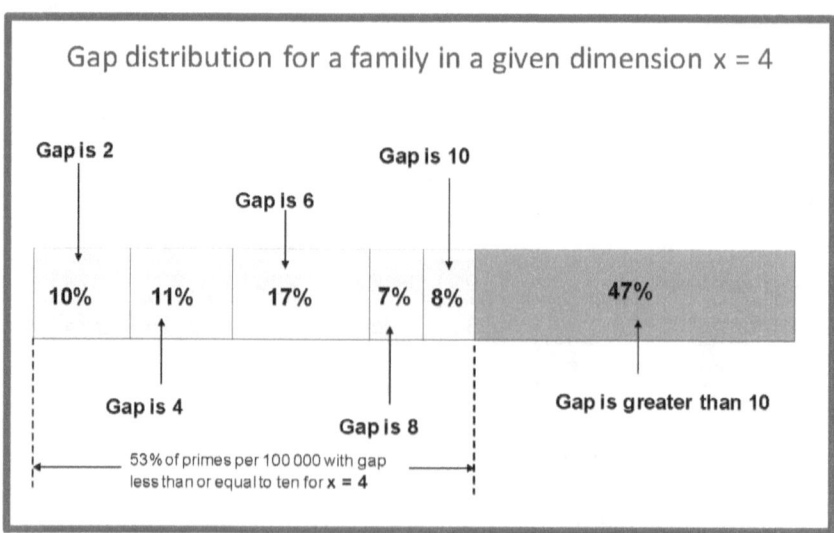

Figure 18. Distribution of gaps within a given prime dimension.

THE THEORY OF PRIME NUMBER CLASSIFICATION

Conjecture 4

There will always be more consecutive primes that have a difference of six than those that have a difference of two in any prime space $P_s = [1, 10^5, s]$.

The basis of the conjecture is found from the percentage of such gaps as shown in table 14, though the table is not regarded as a proof of the theorem. From the table, we can note the behavior of each family across each subspace, and this illustrates a general trend. The gap count for six is consistently above that of two. However, what is more interesting is the fact that the differences less than or equal to ten between consecutive primes is more than 50% for $x = 1$ to $x = 13$.

Conjecture 4

In the prime space $P_s = [1, 10^5, s]$, from $x > 14$, there will always be more primes with a gap greater than ten.

The conjecture implies that for $x > 14$, then $s > 14$ for P_s or for $P_s = [1, 10^5, 14]$, then more primes have a gap less than or equal to ten. In general, the condition "$x > a$" will depend on the depth of the space.

It is demonstrated from the table 14 by the fact that the percentage of prime gaps greater than ten starts to occur from this point onward, and the trend is consistently downward, not upward. While having a random property, the behavioral pattern will be consistent.

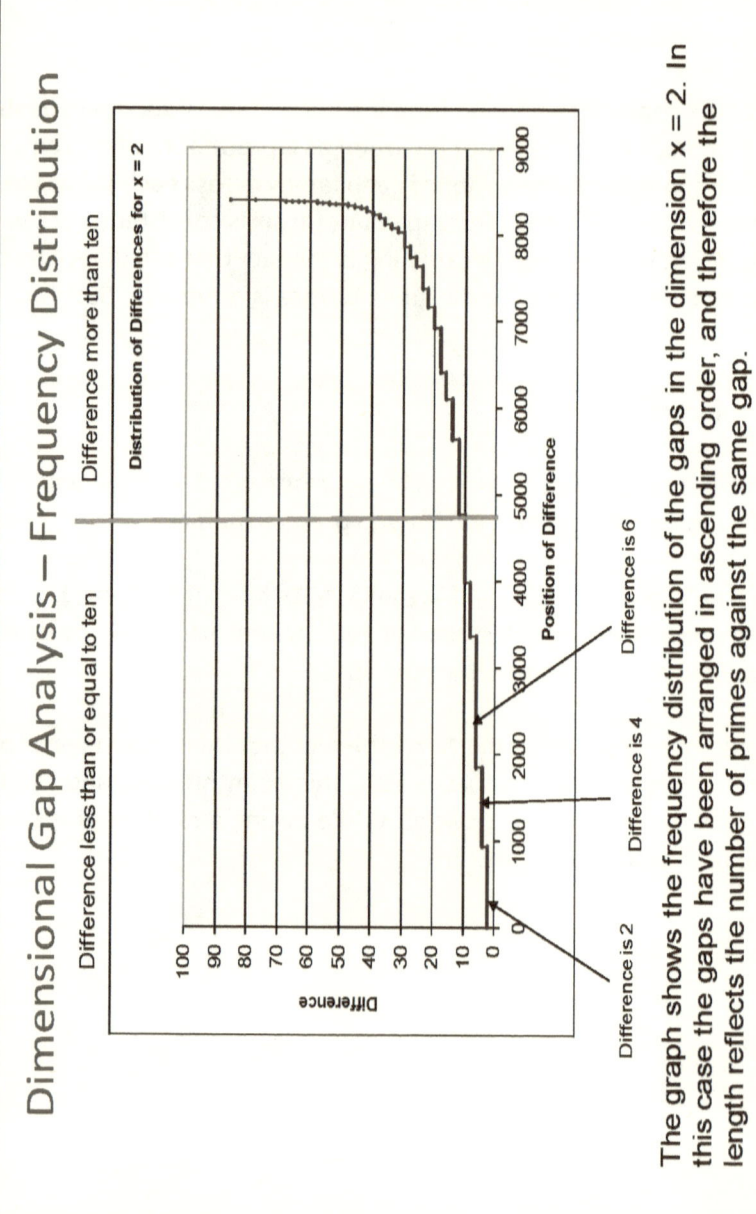

Figure 19. Gap frequency distribution.

THE THEORY OF PRIME NUMBER CLASSIFICATION

x	Gap2	Gap4	Gap6	Gap8	Gap10	Total	Total-2	Gap ≥ 10
1	1227	1216	1941	774	917	6075	9592	
	13%	13%	20%	8%	10%	63%		36.67%
2	937	920	1516	619	790	4782	8392	
	11%	11%	18%	7%	9%	57%		43.02%
3	835	840	1395	591	775	4436	8013	
	10%	10%	17%	7%	10%	55%		44.64%
4	811	839	1342	541	661	4194	7863	
	10%	11%	17%	7%	8%	53%		46.66%
5	762	747	1285	561	704	4059	7678	
	10%	18%	17%	7%	9%	53%		47.13%
6	767	776	1261	499	649	3952	7560	
	10%	10%	17%	7%	9%	52%		47.72%
7	731	752	1203	501	674	3861	7445	
	10%	10%	16%	7%	9%	52%		8.14%
8	706	714	1230	510	654	3814	7408	
	10%	10%	17%	7%	9%	51%		48.52%
9	707	674	1214	514	651	3760	7323	
	10%	9%	17%	7%	9%	51%		48.65%
10	698	674	1172	469	614	3627	7224	
	10%	9%	16%	6%	8%	50%		49.79%
11	726	701	1140	502	608	3677	7216	
	10%	10%	16%	7%	8%	51%		49.04%
12	706	706	1188	493	584	3677	7244	
	10%	10%	16%	7%	8%	51%		49.24%
13	653	667	1118	507	607	3552	7083	
	9%	9%	16%	7%	9%	50%		49.85%
14	702	678	1102	481	577	3540	7105	
	10%	10%	16%	7%	8%	50%		50.18%
15	645	641	1123	476	628	3513	7029	
	9%	9%	16%	7%	9%	50%		50.02%
16	664	657	1083	461	586	3451	6972	
	10%	9%	16%	7%	8%	49%		50.50%
17	671	691	1092	452	570	3476	7014	
	10%	10%	16%	6%	8%	50%		0.44%
18	652	618	1034	432	580	3316	6931	
	9%	9%	15%	6%	8%	48%		2.16%
19	657	636	1069	453	572	3387	6957	
	9%	9%	15%	7%	8%	49%		1.32%
20	631	608	1038	460	581	3318	6904	
	9%	9%	15%	7%	8%	48%		1.94%
21	644	593	1052	431	577	3297	6872	
	9%	9%	15%	6%	8%	48%		2.02%
22	602	606	1092	462	563	3325	6857	

5 SOME APPLICATIONS

x	Gap2	Gap4	Gap6	Gap8	Gap10	Total	Total-2	Gap > 10
	9%	9%	16%	7%	8%	48%		1.51%
23	638	609	1056	423	583	3309	6849	
	9%	9%	15%	6%	9%	48%		1.69%
24	605	592	1031	443	559	3230	6791	
	9%	9%	15%	7%	8%	48%		52.44%
25	557	654	1002	455	528	3196	6770	
	8%	10%	15%	7%	8%	47%		52.79%
26	608	604	1038	448	569	3267	6808	
	9%	9%	15%	7%	8%	48%		52.01%
70	524	506	928	397	521	2876	6367	
	8%	8%	15%	6%	8%	45%		54.83%
100	522	500	832	420	483	2757	6241	
	8%	8%	13%	7%	8%	44%		55.82%
300	454	455	817	342	465	2533	5860	
	8%	8%	14%	6%	8%	43%		56.77%
500	415	394	748	313	411	2281	5594	
	7%	7%	13%	6%	7%	41%		59.22%
800	401	409	677	337	401	2225	5496	
	7%	7%	12%	6%	7%	40%		59.52%
9990	299	292	561	235	321	1708	4760	
	6%	6%	12%	5%	7%	36%		64.12%

Table 14. Group distribution of prime gaps per dimension per hundred thousand.

It is interesting to note that while this was done for primes across families, the same attribute is observed from within a family. For example, consider the graph of the family $x = 2$. If we graph the primes according to their position as the y-axis and the gap as the x-axis, then we derive a curve as shown in figure 17. This is called a gap frequency graph. This is similar for all families, stressing the aspect of making an observation that is true for one instance n and the next family $(n + 1)$, and hence, true for an infinite number of families.

The other observation of interest is the fact that it appears that primes have a repetitive gap pattern based on two, four, and six to define a gap group as shown in the table. The pattern is observable by looking closely at the frequency curve for any given dimension of the prime space. The next "2, 4, 6" is therefore represented by "8, 10, 12" but has a shorter frequency value, and the pattern is repeated along the curve.

Two	Four	Six
8	10	12
14	16	18
20	22	24
26	28	30
..	..	..

Table 15. Gap basic patterns.

The gap group with the largest frequency is the gap variable 6, 12, 18, and so on, whilst the two—and four-gap groups are more or less equal and alternate. However, the six-gap group is consistent in having a larger frequency and so always has a longer length. Two and four have a similar frequency distribution.

5.5 THE CONCEPT OF GAP IN THE GAP

It is common to look at the prime gap in terms of the sequential arrangement of primes, hence the concept of the nth gap. By defining a delta variable, we can also look at a particular gap in isolation and also with respect to other gaps or in comparison to gaps between the prime families. It must be remembered that a delta variable also defines a prime family.

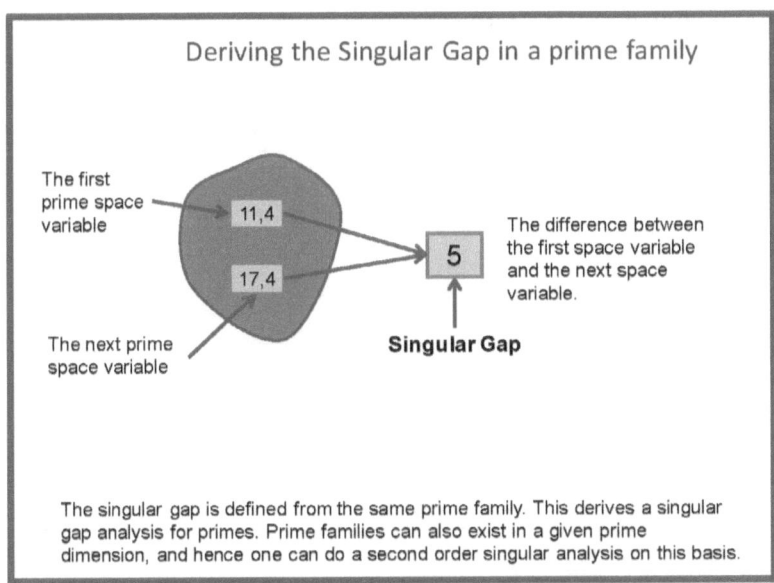

Figure 20. How the singular gap is derived.

Consider, for example, the prime gap 42, the prime numbers being (105319, 42) and (108343, 42). The shadow primes are (105277, 8) and (108301, 8) respectively. In terms of successive primes, the gap is forty-two (delta variable), and the gap between these two is $108343 - 105319 = 3024$, where this defines the "gap in the gap." It is important to note that the first gap is defined for the prime space P_s while the second gap is defined in the space P_f.

Definition 5

For a given prime family, the gap between a prime p and the next prime q is called a singular gap g_s.

This gap is defined in the same conventional approach, the difference being that fact that we are now in the prime space P_s. Therefore,

a. f_g is a gap that defines a prime family in the delta space.
b. g_n is the natural gap of primes in their natural sequential order.
c. g_s is the singular gap of primes in the delta space.

Generally speaking, $f_s = g_s$, but the conceptual roles are different. Hence, just like we may take the average on all the primes in a given subspace x in P_s and find the average gap, a similar exercise is done for the space P_f, where this is now the average of the gap in the gap or a singular gap analysis as it's is called. Continuing our example, the "gap in the gap" differences are a statement of the probability of finding gaps greater than or equal to 42 between a given range of primes. That is, there are 259 primes between 105319 and 108343, and the probability that there is a prime generated with a gap of 42 among them is zero.

THE THEORY OF PRIME NUMBER CLASSIFICATION

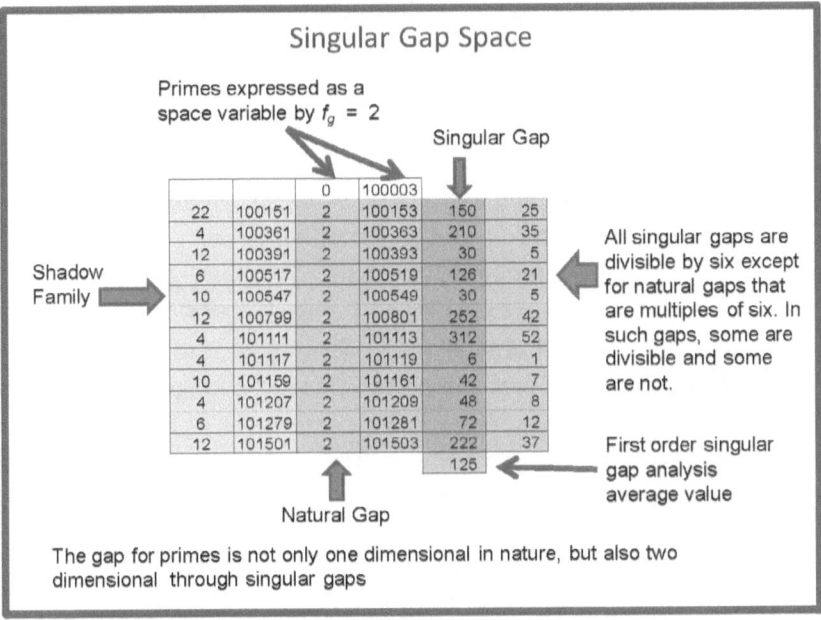

Figure 21. Deriving the singular gap space.

5.5.1 Types of Prime Gap Analysis in P_f

Eight types of prime gap analysis techniques are postulated, where they are called delta analysis.

a. *Between-gap prime count analysis.* This involves taking a given delta variable and finding the number of primes between a given prime and the next prime of the same delta family. This is similar to observing that the gap between the twin prime and the next twin prime grows larger as one goes toward infinity. The number of such primes is actually random and is shown by an example of taking primes in the subspace x = 9, the first thirty primes out of 1213 with gap of six.

b. *Composite gap analysis.* This is the normal gap analysis that consists of finding the average gap of a set of prime numbers in their natural consecutive order in the set P_s. The analysis involves the whole prime subspace for a given x value. The word "natural" is important since in the set P_f, the primes are consecutive but not in their natural order.

c. *First-order singular gap analysis.* For the given subspace x, this involves finding the average of the "gap in gap" in the space P_f. When x is a variable, this plots into a curve, and the analysis is simply called singular

111

analysis. The values differ with those of the composite gap analysis, but the graphs show corresponding behavior. That is, they all increase with x, though the singular gap increases faster. When, for example, we have the analysis of a particular delta variable, say "2," where x is not constant (comparison between families), then this becomes a delta analysis as shown by table 7. Because the delta variable is held constant, this is also referred to as delta analysis.

d. *Second-order singular gap analysis*. This involves finding all the averages of the "gap in the gap" for each delta variable in a given subspace x for the prime space P_f. Therefore, the analysis is vertical since x is constant, and the delta variables change within x and are the object of analysis.

e. *Third-order singular gap analysis*. The analysis consists of comparing the average values of the gap in the gap, where the delta variable is held constant. That is, a given delta variable is held constant while x is a variable in a given space P_f. For example, the delta variable may be "4," then the averages "gap in the gap" for a given range of x is compared.

f. *Comparative gap analysis*. The rate of change of gaps for given variables are compared with the composite gap analysis values.

g. *Proportional gap analysis*. For every subspace, an analysis can be made of the number of delta variables with respect to the total number of primes, where this is expressed as a percentage.

h. *Probability gap analysis*. This is an analysis that finds the probability of finding a gap in a given subspace with respect to a change in x.

In the case of table 16, the singular average involves finding the average gap for the family of "2" only for each x variable. This is a horizontal analysis of the gap distribution with respect to x. The aim of developing such approaches to analyzing prime gaps is to assist in developing a depth in the behavior of primes.

x =	1	2	3	4	5	6	7	8	9	10
Singular Average	82	107	119	123	131	130	137	142	142	143

Table 16. Finding singular average singular gap for gap 2.

THE THEORY OF PRIME NUMBER CLASSIFICATION

x =	11	12	13	14	15	16	17	18	19	20
Singular Average	138	141	153	142	155	151	149	153	152	158

Table 16-B

x =	21	22	23	24	25	26
Singular Average	155	166	157	165	173	164

Table 16-C

Col-1	Col-2	Col-3	Col-4	Col-5	Col-6	Col-7
Prime	Gap	Prime	Delta Variable	Prime	Gap of ga	Acceleration
2	0	2	0	2	0	
3	1	3	1	3	1	
5	2	5	2	5	2	
7	2	7	2	7	2	
11	4	13	2	13	6	
13	2	19	2	19	6	
17	4	31	2	31	12	
19	2	43	2	43	12	
23	4	61	2	61	18	
29	6	73	2	73	12	
31	2	11	4	11		8.75
37	6	17	4	17	6	
41	4	23	4	23	6	
43	2	41	4	41	18	
47	4	47	4	47	6	
53	6	71	4	71	24	
59	6	83	4	83	12	
61	2	29	6	29		10.1428571
67	6	37	6	37	8	
71	4	53	6	53	16	
73	2	59	6	59	6	
79	6	67	6	67	8	
83	4	79	6	79	12	
89	6	89	6	89	10	
97	8	97	8	97	8	
Average	3.8			3.8	9.17391	

Figure 22. Showing classification in terms of gap types.

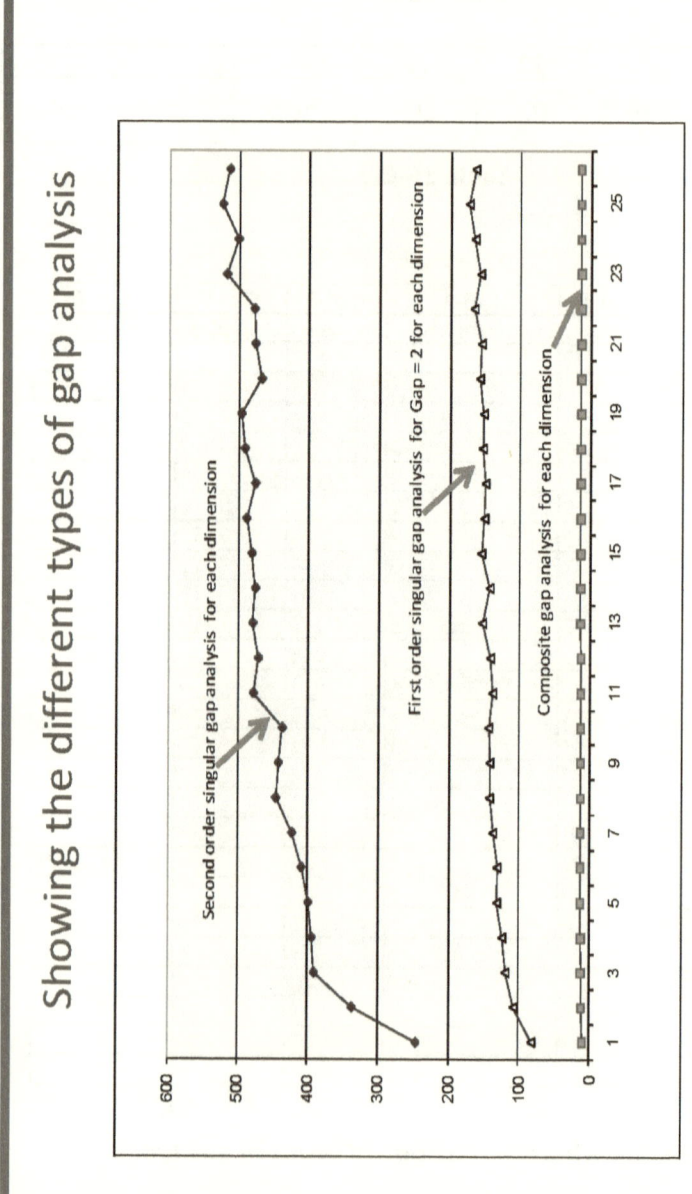

Figure 23. Different types of gap analysis.

All gaps are calculated per dimension. The composite gap is based on difference between the primes. The second singular gap is also a composite of all singular gaps, and then the first order involves looking at the gap for 2 only.

THE THEORY OF PRIME NUMBER CLASSIFICATION

Hence, from figure 22, column 1 consists of such primes, and column 2 consists of the gap between each of the successive primes. As stated before, column 1 therefore represents primes in the prime space P_s. The average gap of such primes is 3.8 as indicated at the bottom of the column, where this is close to the value $Ln(100) = 4.6$. The process of getting the value 3.8 is called a composite gap analysis; it includes all the delta variables of the subspace. The third column represents primes that are arranged according to their gap, which is now referred to as the delta variable space P_f.

In contrast to composite analysis, singular analysis involves, first of all, finding the gap in the gap and then looking for an average. There are different types of singular analysis. The gap in the gap is defined for primes within the same family, that is, those defined by the same delta variable. Looking at figure 22, the value 9.17 is the average of "gap in gap" for the subspace and the process is called the first-order singular analysis. Therefore, the first order is for the whole subspace or group while the second order is for each of the subgroups defined by a delta variable. This is shown by the gap calculated in the seventh column. The average for the delta variable "2" is 8.75 and 12 for the delta variable "4."

The process of deriving this value is called the second-order singular analysis. However, when we repeat the same analysis with "2" in another subspace of P_s, this now involves a family of primes defined by the gap "2," and this becomes a third-order singular analysis. Hence, a delta analysis is horizontal involving the same family whilst the second-order singular analysis is vertical in the same subspace. Therefore, the value average 8.75 and 12 for delta variables "2" and "4" are called the gap acceleration of the primes in that family—these represent an acceleration of a particular delta group.

Definition 6

Prime acceleration is defined as the average rate of gap change of the gap in the gap of a prime family or subspace in the prime space P_f.

Consequently, there are two ways of studying the average gap in the gap whole of each subspace in a given prime space, that is, for the given span of the space. This defines a general trend of the average of the gap in the gap for each subspace and forms first-order singular analysis. For example, consider the average gap in

5 SOME APPLICATIONS

the gap for the space $P_s = [1, 10^5, 30]$, then we see that this average is actually increasing as expected with x as shown in figure24.

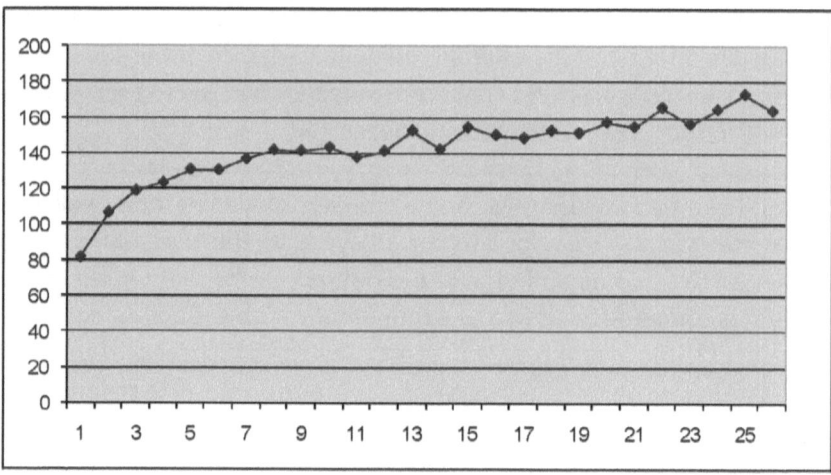

Figure 24. The average singular gap for a given space [1, 10^5, 30].

Consider the average gap in the gap for the variable "2." This curve is increasing with x, which implies that the gap of the gap for two is widening as x increases. A similar exercise is done for the family defined by the delta variable "42" for each subspace. This is like taking a microscopic view, giving us more detail then the general trend. On the other hand, for the delta variable "24," it is said to be steady as shown in figure 25-27.

Definition 7

For a given family of primes, the rate of change of the acceleration is defined to be either positive, negative, or steady.

Consequently, if the acceleration increases with x, then it is said to be positive, while if it decreases with x, it is said to be negative. Hence, for the delta variable 2, the acceleration of the primes is positive, that is, the gap of the gap increases with x. In fact, there are three possible outcomes for acceleration that are assumed, that is,

- acceleration can be positive, as in the delta variable 2,

THE THEORY OF PRIME NUMBER CLASSIFICATION

- acceleration can be negative, as in the delta variable 42,
- acceleration can be constant or steady, as in the delta variable 24.

In order to create an interesting picture, we can put all of these in one graph, illustrate the concept for the average gap for each x and a singular gap analysis for x and the delta-two variable gap values. The singular graph is the acceleration of the gap of the prime numbers in each subspace, and it is increasing with x. This means as a whole, looking at the subspaces, then there is a general widening of the gaps within the gaps themselves. The composite value is the gap value that is an average in P_s, it is not concerned with the gap that is a consequence of arrangement of primes according to their gap differences. This value is much lower. The delta variable gap value is higher but is lower than the singular value.

This is shown by figure 23. So it may be noticed that the composite gap does not express this information, but this is typical of looking for averages anyway. Some information become lost unless other methods of looking for it are defined. Similarly, the singular gap does not demonstrate these differences in the way the primes behave until we apply the gap analysis family by family.

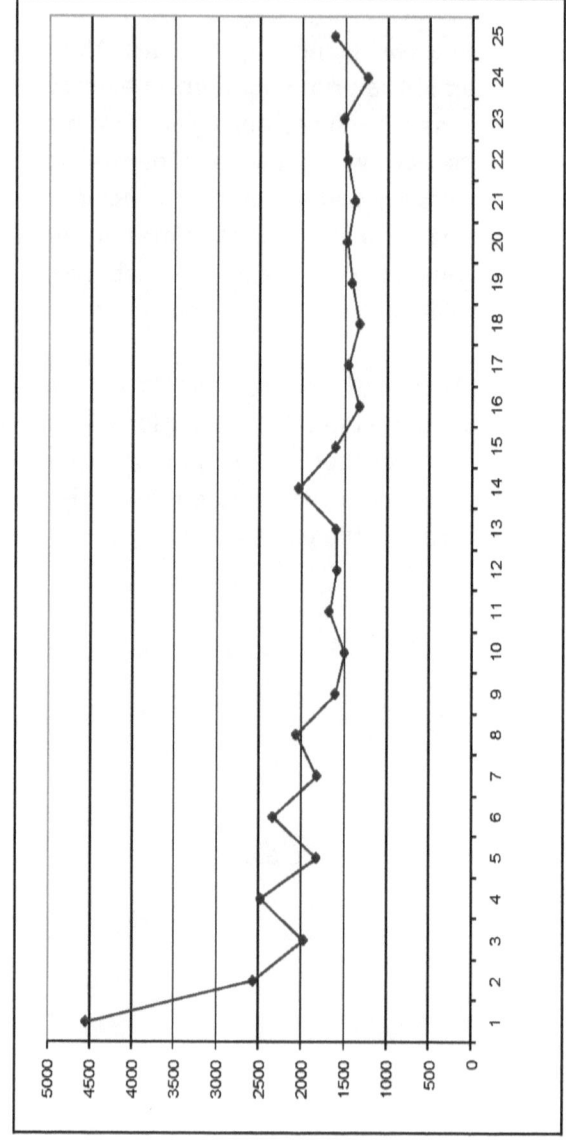

The graph shows the rate of change of the gap 42 with respect to the prime dimension.

Figure 25

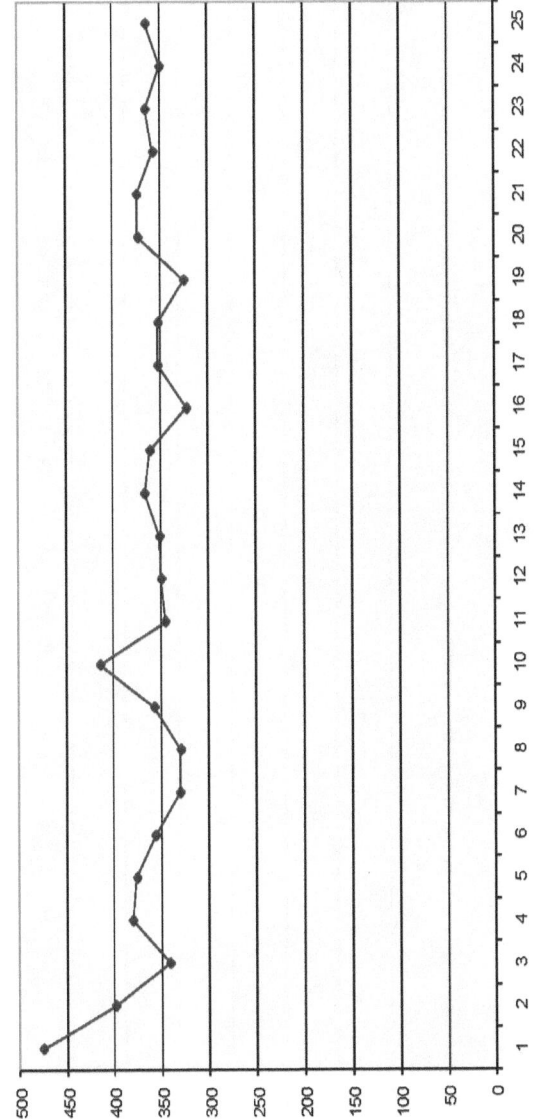

Figure 26

The graph shows the rate of change of the gap 24 with respect to the prime dimension.

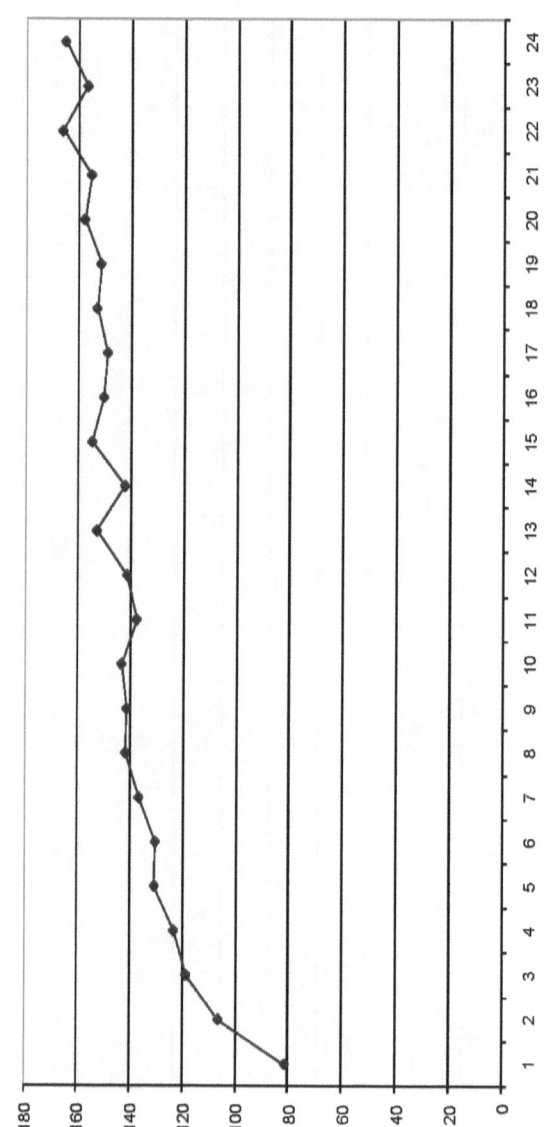

Figure 27

The graph shows the rate of change of the gap 2 with respect to the prime dimension.

Table 17 further shows the relationship between the accelerations of gaps. The table shows the values of acceleration for each gap (vertical analysis) and also for each x (horizontal analysis). The subspaces under consideration are $x = 1, x = 5, x = 25$, and $x = 9990$. The analysis is conducted for the prime space P_f. The gaps for which the acceleration is stated as unclassified mean no verification has been made as to whether the acceleration is increasing or decreasing. They appear to be steady, but more values need to be investigated to confirm the actual behavior. The table also shows the average acceleration for prime gaps less than or equal to ten and the average for more than ten but less than or equal to sixty. The last column indicates the type of change that is observed for the given gap, where this can be positive, steady, or negative. It is of further interest to note that these graphs help us to interpret and explain the table 14, where it was showing how the proportional number of primes per hundred thousand whose gap was greater than ten was increasing, while those whose gap was less than ten was decreasing. Using this information of acceleration, primes less than ten accelerate much slower, even though they has a positive acceleration, while those with a higher gap have a negative acceleration but accelerate much faster. Hence, table 18 gives us a clue of a net difference of the two accelerations and shows the consequence result of the two accelerations. For primes less than or equal to ten, the acceleration is increasing, which implies the gap between the primes is increasing. For primes greater than ten, the prime acceleration is decreasing, which implies that the gap between primes is reducing. Hence, there will be less primes comparatively with a gap less than or equal to ten and more primes with a gap more than ten.

5.5.2 Comparing Gap Accelerations

We can make simple gap comparisons between prime families or between singular gap values. This is done by looking for the trend average in each situation. For example, let us consider finding out how much faster is the "2"

Gap	x = 1	x = 5	x = 25	x = 9990	Type of Change
2	82	130	164	334	Positive
4	82	129	165	341	Positive
6	52	79	96	178	Positive
8	129	200	223	425	Positive

10	109	154	176	312	Positive
Average	91	139	165	318	Positive

Table 17. Prime acceleration for gap less or equal to ten.

family gap increasing with respect to the composite average of the prime gap. For the composite gap average, it is,

$$y_c = 1.1915Ln(x) + 10.965 \quad (1)$$

For the delta-two variable, it is,

$$y_t = 23.045Ln(x) + 89.074 \quad (2)$$

The intercepts already tell part of the story—that the gap is higher for the delta-two variable. Hence, we have,

$$y_c - 10.965 = 1.1915Ln(x) \quad (3)$$

$$y_t - 89.074 = 23.045Ln(x) \quad (4)$$

Now we want to compare the values of y_c and y_t for the same value of x; hence, $L(x)$ will cancel if we divide by the top equation. That is,

$$y_t - 89.074 = 19.3412(y_c - 10.965) \quad (5)$$

Or

$$y_t = 19.341(y_c - 10.965) + 89.074 \quad (6)$$

Gap	x=1	x=5	x=25	x=9990	Change
12	103	128	147	246	Positive
14	206	249	269	454	Positive
16	290	363	381	580	Positive
18	194	197	220	335	Positive
20	418	464	448	542	Unclassified
22	443	459	504	703	Positive
24	476	357	365	476	Constant

26	1116	709	851	889	Unclassified
28	998	789	696	865	Unclassified
30	656	480	485	508	Unclassified
32	3043	1756	1259	1528	Unclassified
34	2991	2448	1801	1230	Negative
36	1683	1658	911	759	Negative
38	3624	2498	2503	1874	Positive
40	2889	1972	1824	1548	Negative
42	4533	2332	1619	1009	Negative
44	19206	4058	3824	3222	Negative
46	4058	4569	4127	3275	Unclassified
48	29200	3558	2841	1612	Negative
50	15936	8131	5421	2424	Negative
52	9225	9881	7833	4203	Negative
54	17084	7287	3538	2066	Negative
56	—	9541	13068	4034	Negative
58	—	15908	8343	3540	Negative
60	—	5611	4678	2108	Negative
Average	5380	3416	2718	1601	Negative

Table 18. Prime gap acceleration for gap greater than ten.

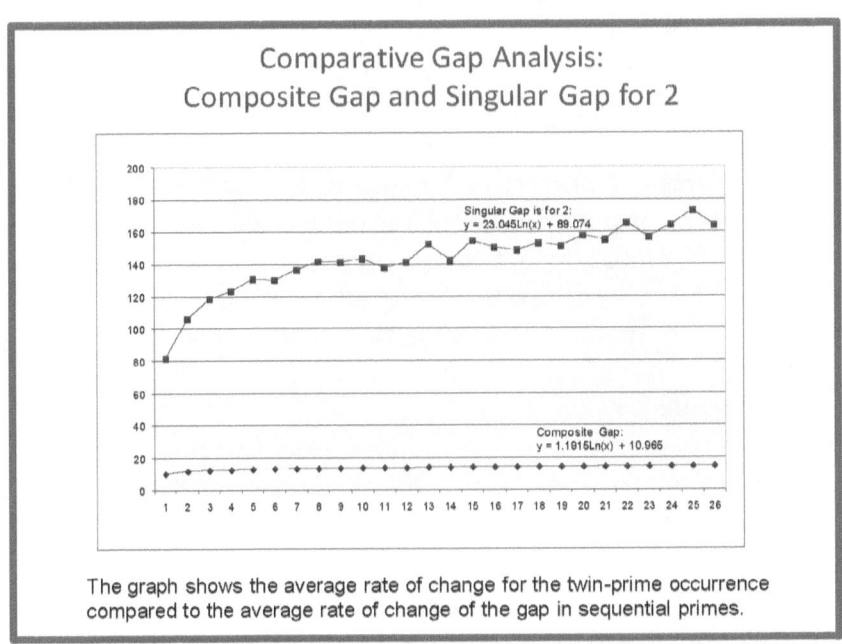

The graph shows the average rate of change for the twin-prime occurrence compared to the average rate of change of the gap in sequential primes.

Figure 28. Comparative analysis.

5 SOME APPLICATIONS

Our interest lies in the gradient value 19.34, as it gives us an approximation of the rate of change of the delta two variable with respect to the composite gap value. From the above, we can then state that the gap "2" in the primes is increasing at approximately nineteen times faster than the average rate of the primes in their consecutive order. A similar exercise may be done for the singular gaps, where comparison may be between,

a. two singular gaps,
b. two delta gaps, and
c. a composite gap with a delta or singular gap.

Finding the approximate rate of increase of one with respect to the other gives us an idea of the behavior of the primes. For example, the delta six variable increases at a slower rate that the delta two variable in terms of gap acceleration as well as the gradient comparison.

5.5.3 PROBABILISTIC GAP ANALYSIS

A proportional analysis picks a certain range of primes and determines their relative distribution percentage per hundred thousand in terms of the delta variable. Hence, for example, if we take the group of primes at $x = 4$ in a prime space $[1, 10^5, 30]$, then as shown in figure 17, the proportion of gaps with respect to each other is 10%, 11%, 17%, 7%, 8% for gaps 2, 4, 6, 8, and 10 respectively. That is, 53% of the primes have a gap equal to or less than ten as observed from table 17. Therefore, this gives us a proportional distribution of the primes in the given prime space. This can be also expressed as a probabilistic event, that is, the probability of finding a prime gap that is ten or less is more than half that is $0 \cdot 53$. For $x = 9990$, the probability of finding a gap that is ten or less is $0 \cdot 36$. Therefore, the following conjecture may be drawn.

Conjecture 5

The higher the value of x, then the higher is the probability of finding a gap that is greater than ten.

The implication of the conjecture is that the population of smaller gaps decreases probabilistically such that there is a form of displacement by the family of larger gaps. That is, there is a definite relationship between the behavior of smaller gaps and larger gaps and their existence in a given prime price that is inversely related.

Theorem

For a given prime subspace, the probability of smaller gaps is inversely related to the probability of larger gaps.

Proof. A given prime space x is bounded; it has an initial prime and an end prime. Between the initial prime and the end prime, there must be n primes with some gaps in between, where such gaps can either be $2, 4, 6, 10 \ldots, 2k$ and so on up to infinity.

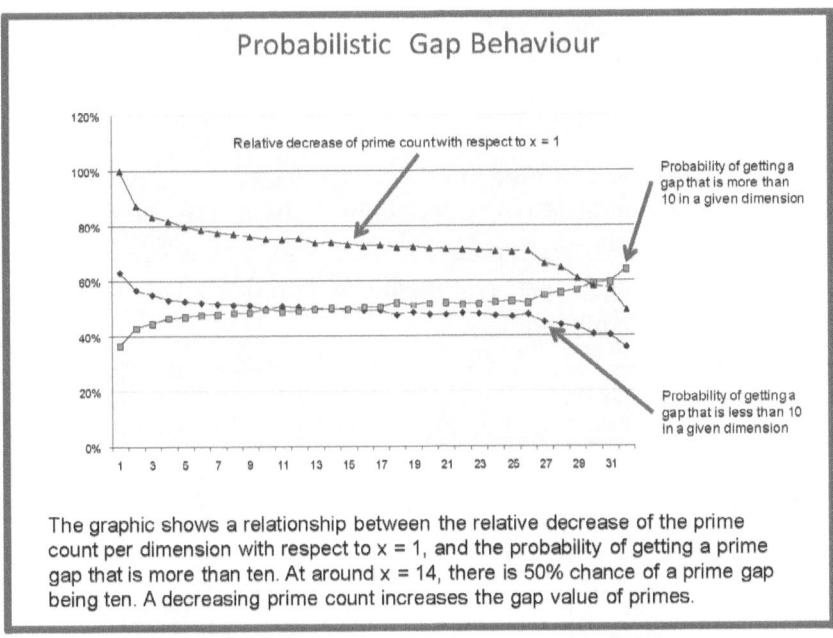

The graphic shows a relationship between the relative decrease of the prime count per dimension with respect to x = 1, and the probability of getting a prime gap that is more than ten. At around x = 14, there is 50% chance of a prime gap being ten. A decreasing prime count increases the gap value of primes.

Figure 29. Probabilistic relation of prime gaps.

Hence, if it is assumed that the probability of finding a gap less than ten in a space x is $P(a)$, then the probability of finding a gap greater than ten will be $1 - P(a)$. Therefore, if $P(a)$ is constant, then $1 - P(a)$ will also be constant. If $P(a)$ increases, consequently $1 - P(a)$ will decrease, or if $P(a)$ decreases, then $1 - P(a)$ will increase. Hence, in the given space x, the relation will be inversely related when x changes $1 - P(a)$ either increases of decreases. Since

5 SOME APPLICATIONS

the number of primes decreases asymptotically, then $P(a)$ cannot be constant; hence, the relation is inversely proportional. [End Proof].

This is also demonstrated by figure 30.

Therefore, if we look at $x = 8$ for example, we note that there are 7408 primes in this subspace. Of these primes, 3814 are primes that have gaps less than or equal than to ten. This means 51% of the primes in this subspace have a gap less than or equal to ten, that is $P(a) = 0 \cdot 51$. Therefore, the following can be noticed. Observing from figure 28, it shows that as the number of primes decrease in x probabilistically, we should expect the gap in x to increase probabilistically (see top graph in figure). The first subspace $x = 1$ has 9592 primes, and so if this is taken as 100%, a comparative analysis of decrease in the prime count per hundred thousand can be made. This gives us a curve that has a downward trend, describing the probability of finding a prime subspace that has the same number of primes as $x = 1$ as x increases.

A similar example is done for the interval of 100 for each subspace in order to establish the distribution of prime occurrence. The results are obtained and shown by table 10. Zero means no occurrence of a prime in a given interval of 100, one means an occurrence, two means two primes in the 100 space, and so on. See illustration in figure 29.

x =	0	1	2	3	4
1	26%	40%	23%	9%	2%
2	34%	40%	20%	5%	0%
3	40%	40%	16%	3%	1%
4	39%	42%	15%	3%	0%
5	44%	38%	14%	3%	0%
6	42%	42%	14%	2%	0%
7	44%	40%	14%	2%	0%
8	47%	38%	12%	3%	0%
70	57%	35%	7%	1%	0%
800	65%	29%	5%	0%	0%
9990	73%	24%	3%	0%	0%

Table 19. Space 100 probability distribution.

THE THEORY OF PRIME NUMBER CLASSIFICATION

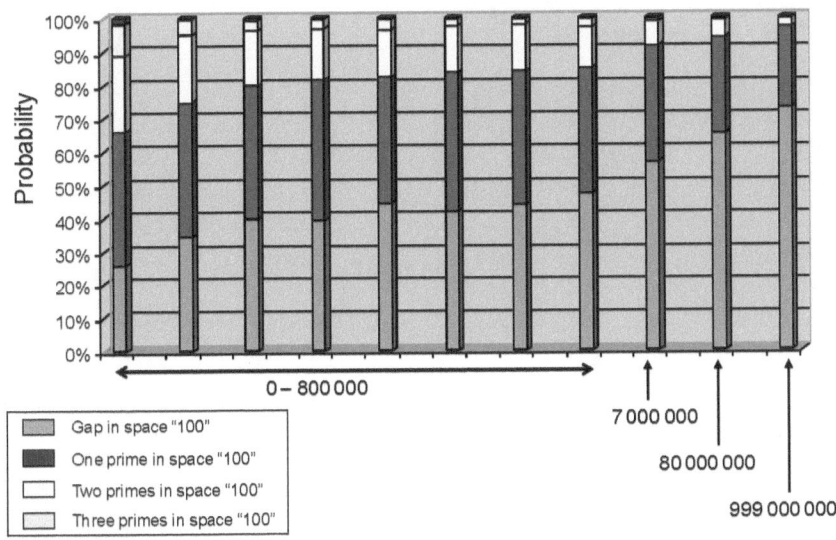

Figure 30. Probability distribution for 100 interval.

Section Four

Gap Theory Classification of Prime Numbers

Overview

Interesting gap patterns can be studied in prime number theory, where this can occur as a group of primes having a certain gap or common characteristic. The aim here is to use gaps as one way of classifying prime numbers in general. The first approach used prime roots. This may contribute to understanding the predictive behavior of prime numbers.

Objective:

To study the predictive and linear behavior of primes numbers in a two-dimensional space.

Research Question

What is the influence of the prime gap on predicting the position of a given prime number in the prime space?

1 Introduction

The study of prime numbers centers a lot on their density in terms of distribution and also on the gaps between primes. The major focus on these is due to the fact that there is an assumption that they probably hold the key to the predictability of occurrence of the prime numbers. A lot of deep analytical research has gone into prime numbers and the type of patterns that they form, and several conjectures and theorems still remain to be proved. The basic framework of gap analysis is based on the prime number theorem, which is also used as a tool to describe the asymptotic distribution of the prime numbers. The theorem states that the average length of the gap between a prime p and the next prime is $Ln(p)$, where the actual length of such a gap might be much more or less than this.

Whilst there are many formulas to describe certain primes, famous amongst them being Mersenne primes and Fermet primes, the simple formula that any prime can be written in the form $6n \pm 1$ is often ignored or overlooked leading to the situation that many lone rangers discover it and make a proclamation. It is ignored because it has no predictive powers for prime numbers and does not seem to offer anything else besides the fact that any prime can be described using this equation, which can also be done for example by the equation $2n + 1$.

2 Gap Theory of Prime Number Classification

Any natural number can be written in the form $6a \pm b$, where a is any integer, and $b \in H = \{0, 1, 2, 3, 4, 5\}$. If we consider each value of b individually, then,

a. when $b = 0$, then $6a \pm b = 6a$, and this is divisible by six, so it can never represent a prime number.
b. when $b = 1$, then from $6a \pm 1$ there is a possibility for a prime number; for example in $a = 3, b = 1$, then $6(3) + 1 = 19$, a prime. The same applies for $b = 5$, that is $6a \pm 5$, since for example we can have $6(3) + 5 = 23$ and $6(3) - 5 = 13$.
c. when $b = 2, 3, 4$, this gives us $6a + 2 = 2(3a + 1)$, and $6a + 3 = 3(2a + 1)$, and $6a + 4 = 2(3 + 2)$ respectively, and these do not define prime numbers.

Even though only two conditions based on divisibility are enough to define the prime number, we can actually define a prime as having three characteristics so that we understand its context in the natural numbers. The third characteristic is the gap. This then allows us to develop the following axioms with regard to the prime number and gaps. The focus of the axioms is the resulting gaps, not the primes that are now taken as the cause. That is, one may also consider prime numbers as a result of gap behavior rather than the other way round.

Prime Gap Axioms

a. Any prime p number satisfies $p \in \{2, 3, 6a \pm 1\}$ and $a \in N$, the set of natural numbers.
b. For $6a \pm 1$, let $a = f(x_a) + c$, where c is some constant.

c. For the prime space P_s, let there be two consecutive primes p_n and p_{n+1}, then define the sequential prime gap $g_n = p_{n+1} - p_n$ as the natural gap.
d. For the prime space P_f, then $g_n = f_g$, where f_g defines a prime family (p_x, f_g).
e. For a prime family (p_x, f_g), let there be two consecutive primes p_1 and p_2, then $g_s = p_2 - p_1$, where g_s is called the singular gap.
f. Let $(p_2, f_g), (p_4, f_g), (p_6, f_g)$ be the base gaps of the delta space, then this defines classification group 1, group 2, and group 3 of primes respectively.
g. Prime numbers in a given family of gap f_g have a linear relationship.

Probably the most important condition is the last one. It is based on the observation and assumption derived as follows. For a given family (p_x, f_g), let $n(p_x, f_g) = h$ define the number of elements in the subset of primes. Hence, the format $y = mx + c$ must be satisfied. Then a prime p_t occuring at t, we have,

$$f(t, p_t) = m_x t + k$$

for $1 \leq t \leq h$, k is constant, m_x is a gradient for dimension x. This equation defines a prime number spectral line, where this is to be demonstrated later. This agrees with the basic axiomatic assumption that all prime numbers satisfy the relation that $p \in \{2, 3, 6a \pm 1\}$.

What is significant about the axioms is the fact that they set the framework for defining another way of classifying the prime numbers. This is called the gap theory of prime number classification. The system is constructed on the basis of the gap 2, 4, and 6 called the base gaps.

Definition 1

Let gap 2, gap 4, gap 6 construct a gap-base system for classification and generation of any type of prime numbers.

This is a far-reaching definition, though simple in form. If we construct prime numbers in terms of a gap-base system, then it also changes the concept of the

2 GAP THEORY OF PRIME NUMBER CLASSIFICATION

definition of a prime number and has the implication of producing different types of prime numbers. The definition also implies that there is a specific system that describes and defines the prime number, hence the notions of classification and generation.

Now gap 2 goes along with $8, 14, 20, ...$, gap 4 with $10, 16, 22,...$, and gap 6 with $12, 18, 24, ...$, where each goes to infinity. The first point we note on this system is the following:

> *The difference between each member of the same set of gaps is six.*

Secondly, each horizontal group of primes for the given gap can then be called a class while each gap base defines a prime number category.

Gap Category	6a+1 Divisible by 6	6a-1 Divisible by 6	(6a+1) or (6a-1) Not always multiple of 6
Class 1	2	4	6
Class 2	8	10	12
Class 3	14	16	18
Class 4	20	22	24
Class 5	26	28	30
Class n	...	...	...

Figure 31. Showing the gap classes and category.

An interesting observation then emerges given by the following definition:

Definition

a. *All primes of gap base 2 are defined by (6a+1).*
b. *All primes of gap base 4 are defined by (6a-1).*
c. *All primes of gap base 6 are defined by either (6a+1) or (6a-1).*

Furthermore, all singular gaps in gap base 2 and 4 are divisible by six with no remainder. In other words, $p \in \{2, 3, 6a \pm 1\}$ is a significant axiomatic condition to describe behavior of the prime numbers.

Therefore, suppose in (p_x, f_g), we want to find the next prime p_2 from some p_1, then let

$$p_1 = 6a + 1$$

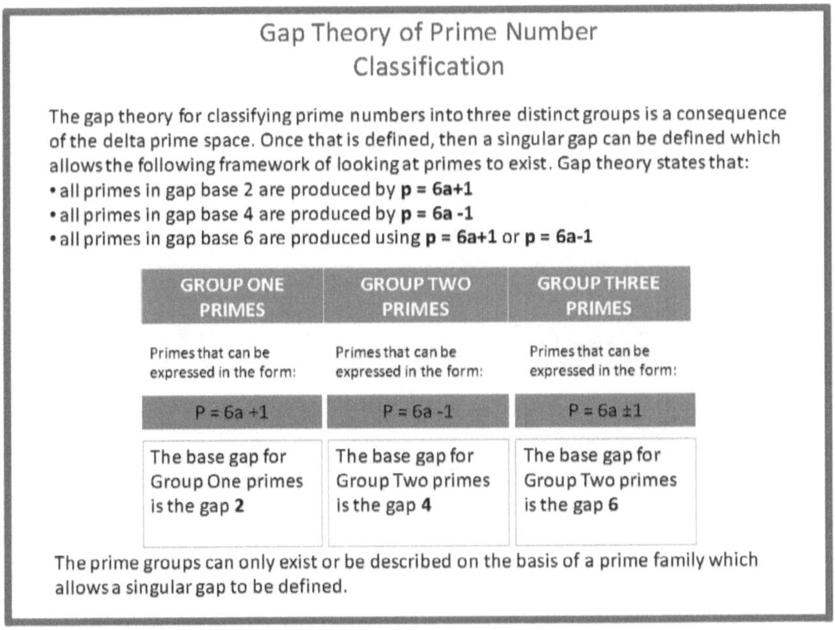

Gap Theory of Prime Number Classification

The gap theory for classifying prime numbers into three distinct groups is a consequence of the delta prime space. Once that is defined, then a singular gap can be defined which allows the following framework of looking at primes to exist. Gap theory states that:
- all primes in gap base 2 are produced by **p = 6a+1**
- all primes in gap base 4 are produced by **p = 6a -1**
- all primes in gap base 6 are produced using **p = 6a+1** or **p = 6a-1**

GROUP ONE PRIMES	GROUP TWO PRIMES	GROUP THREE PRIMES
Primes that can be expressed in the form:	Primes that can be expressed in the form:	Primes that can be expressed in the form:
P = 6a +1	P = 6a -1	P = 6a ±1
The base gap for Group One primes is the gap **2**	The base gap for Group Two primes is the gap **4**	The base gap for Group Two primes is the gap **6**

The prime groups can only exist or be described on the basis of a prime family which allows a singular gap to be defined.

Figure 32. Classification of prime numbers using gap theory.

$$p_2 = 6b + 1$$

Then,

$$p_2 - p_1 = 6(b - a)$$

If $b - a = q_1$, this simplifies to,

$$p_2 - p_1 = 6q_1$$

2 GAP THEORY OF PRIME NUMBER CLASSIFICATION

In a similar manner, if we subtract the following:

$$p_2 = 6b - 1$$

$$p_1 = 6a - 1$$

We get the same result $p_2 - p_1 = 6q_1$. Now $6q_1$ is called the singular gap of the delta space and is true for any two primes in gap base 2 and gap base 4. We can also have,

$$p_1 = 6a + 1$$

$$p_2 = 6b - 1$$

Now subtracting, this yields,

$$p_2 - p_1 = 6q_1 - 2$$

Since $b - a = q_1$. For

$$p_1 = 6a - 1$$

$$p_2 = 6b + 1$$

We get

$$p_2 - p_1 = 6q_1 + 2$$

In general, we can write,

$$p_2 - p_1 = 6q_1 + R, where\ R = 0\ or\ R = 2$$

Similarly, we have,

$$p_3 - p_2 = 6q_2$$

Since $p_2 = 6q_1 + p_1$, then,

$$p_3 = 6q_1 + 6q_2 + p_1$$

THE THEORY OF PRIME NUMBER CLASSIFICATION

In general, this gives for the nth prime in a family:

$$p_n = 6(q_1 + q_2 + \ldots + q_n) + p_1$$

Hence, if we can establish the gap sequence q_i for group 1 and 2 primes, we can then predict the next prime.

3 Structure of Gap Relationships for Classification

Hence, the gap theory of classification is based on using the relationships expressed by the equations:

a. For gap base 2: $6a + 1$. This means a prime is produced by this equation; hence, we can determine if the equation was used by an inverse operation given by $(prime - 1)/6$.
b. For gap base 4: $6a - 1$. This means a prime is produced by this equation; hence, we can determine if the equation was used by an inverse operation given by $(prime + 1)/6$.
c. Gap base 6: $6a + 1$ or $6a - 1$. This means a prime is produced by one of these two equations. Hence, we can determine which equation was used by an inverse operation applying $(prime - 1)/6$ or $(prime + 1)/6$ respectively. Therefore, this gives us gap base 6-A and gap base 6-B.
d. We apply the rule that $a \in N$, the set of natural numbers (positive integers).

Now prime numbers in a family are related by the singular gap such that the next prime is a function of this. That is, this relationship is defined as follows:

a. Gap base 2: The singular gap for all primes is divisible by 6
b. Gap base 4: The singular gap for all primes is divisible by 6
c. Gap base 6: The singular gap here can be produced by

  I. $(6a + 1) - (6b + 1)$, and all of these are divisible by 6
  II. $(6a - 1) - (6b - 1)$, and all these are divisible by 6

III. $(6a + 1) - (6b - 1)$, and this is divisible by $(g_s + 2)/6$, where $g_s = p_2 - p_1$ is the singular gap. That is, we have $g_s = 6q + 2$.

IV. $(6a - 1) - (6b + 1)$, and this is divisible by $(g_s - 2)/6$, where $g_s = p_2 - p_1$ is the singular gap. That is, we have $g_s = 6q - 2$.

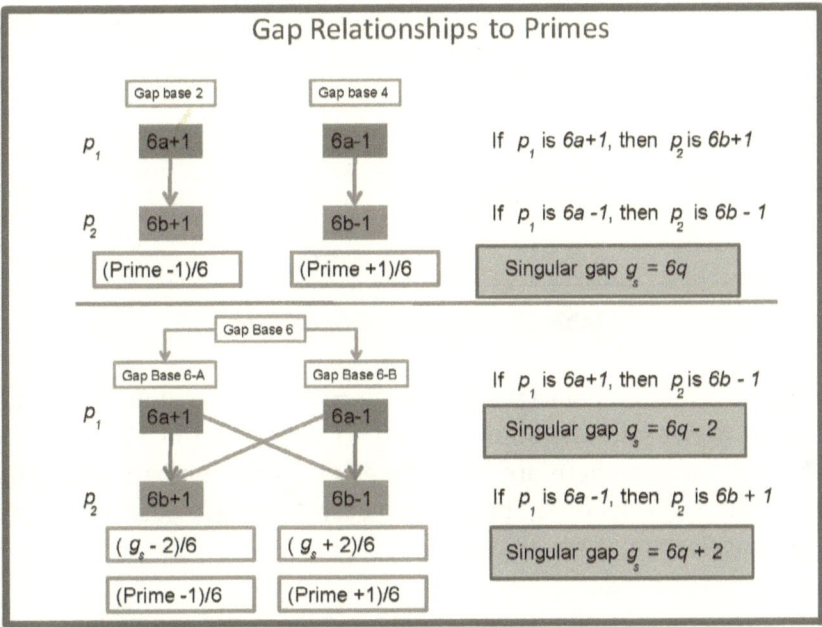

Figure 33. Structure of classification of prime numbers.

We can then group these formulas in the way they influence the distribution and occurrence of the primes.

Classification	Gap Base Two	Gap Base Four
PURE GROUP OF PRIMES	$p_1 = 6a + 1$ $p_2 = 6b + 1$ $q_1 = b - a$ $p_2 - p_1 = 6q_1$	$p_1 = 6a - 1$ $p_2 = 6b - 1$ $q_1 = b - a$ $p_2 - p_1 = 6q_1$

Table 20. Gap base 2 and 4 equations.

3 STRUCTURE OF GAP RELATIONSHIPS FOR CLASSIFICATION

For gap base 6, we have the following distribution and occurrence pattern.

Classification	Gap Base Six-A	Gap Base Six-B
MIXED GROUP OF PRIMES	$p_1 = 6a + 1$ $p_2 = 6b + 1$ $q_1 = b - a$ $p_2 - p_1 = 6q_1$	$p_1 = 6a - 1$ $p_2 = 6b - 1$ $q_1 = b - a$ $p_2 - p_1 = 6q_1$
CROSSING GROUP OF PRIMES	$p_1 = 6a + 1$ $p_2 = 6b - 1$ $q_1 = b - a$ $p_2 - p_1 = 6q_1 - 2$	$p_1 = 6a - 1$ $p_2 = 6b + 1$ $q_1 = b - a$ $p_2 - p_1 = 6q_1 + 2$

Table 21. Gap base 6 equations.

We note the common characteristic $q_1 = (b - a)$ in all the prime descriptions. This gives a two-dimensional method to describe the prime number distribution. That is, there are three types of prime numbers according to the gap theory. These are gap base 2, gap base 4, and gap base 6. In terms of further classification, there are the,

a. pure group of primes whose singular gap is defined by (gap 2, $6q_1$) and (gap 4, $6q_1$);
b. mixed group of primes whose singular gap is defined by (gap 6-A, $6q_1$) and (gap 6-B, $6q_1$); and
c. crossing group of primes whose singular gap is also defined by the (gap 6-A, $6q_1 - 2$) and (gap 6-B, $6q_1 + 2$) variables.

Group Type	Gap Base	Prime Type	Group Factor - R	Singular Gap Type	Grouping Formula	Gap Formula
Pure Group	2	6a+1	R = 0	Singular = Gap + R	(prime − 1)/6	Gap/6
Pure Group	4	6a−1	R = 0	Singular = Gap + R	(prime +1)/6	Gap/6
Mixed Group	6-A	6a+1	R = 0	Singular = Gap + R	(prime− 1)/6	Gap/6
Mixed Group	6-B	6a−1	R = 0	Singular = Gap + R	(prime + 1)/6	Gap/6
Crossing Group	6-A	6a+1	R = 2	Singular = Gap - R	(prime + 1)/6	(Gap-2)/6
Crossing Group	6-B	6a−1	R = 2	Singular = Gap + R	(prime - 1)/6	(Gap+2)/6

Figure 34. Classification framework.

THE THEORY OF PRIME NUMBER CLASSIFICATION

This is best illustrated by example.

Example

Consider the following prime numbers in the dimension x = 999 and $f_g = 6$.

	Family	Prime Number	Gap	(Gap + R)/6	6a ±1
Line 1	6	999032317			166505386
Line 2	6	999032549	232	39	166505425
Line 3	6	999032627	78	13	166505438
Line 4	6	999033487	860	143	166505581
Line 5	6	999033619	132	22	166505603

Figure 35. Primes in dimension x = 999

Line 1

Since this is the initial prime number for this group, we can only classify it. We classify it using the expression $p_1 = 6a \pm 1$.

$$6a + 1 = 999032317$$

$$6a - 1 = 999032317$$

Hence, if we solve, then $a = 166505386$ or $a = 166505386.3$, that is, we use $6a + 1$. The rule states that $a \in N$, it must be an integer; therefore, the prime number belongs to gap base 6-A. There is no gap against this prime number as it is the initial one; hence, we cannot tell whether it is a mixed or crossing prime.

Line 2

The number 999032549 is equated to $6a - 1$. The number is a crossing prime since the previous number divided by $6a + 1$. The singular gap between the two primes is $gap = 230$, and hence, we use $(g_s + 2)/6$, that is,

$$q_1 = \frac{Gap + 2}{6} = 39$$

3 STRUCTURE OF GAP RELATIONSHIPS FOR CLASSIFICATION

Line 3

The number 999032627 is also classified by $6a - 1$, which means it belongs to the mixed group 6-B. In this case, the singular gap between the two primes is $gap = 78$ and $R = 0$. Hence, $q_1 = 13$.

Line 4

The prime number 999033487 is classified by $6a + 1$, and it is a crossing prime because the previous prime was classified by $6a - 1$. This means it belongs to crossing group 6-A. Here, $R = 2$, so $gap = 860 - 2 = 858$. That is, the singular gap is 858. Hence, $q_1 = 143$.

Line 5

The prime 999033619 is classified by $6a + 1$, so this is a mixed prime since the previous was also classified by $6a + 1$. Here, $R = 0$, singular gap = 132, and $q_1 = 22$.

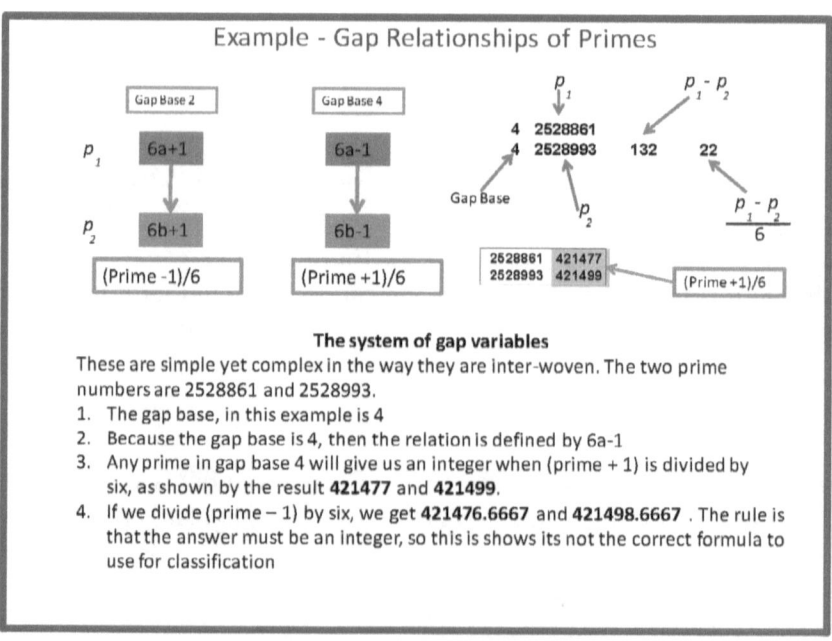

Figure 36. Example in gap base 4.

THE THEORY OF PRIME NUMBER CLASSIFICATION

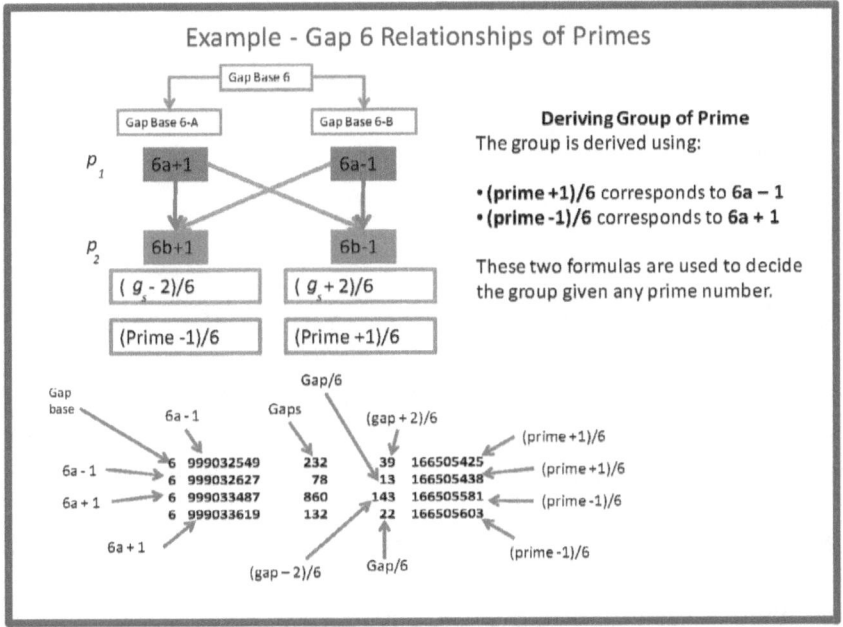

Figure 37. Example in gap base 6.

4 Fundamental Gap Behavior—Spectral Lines

Consider a two-dimensional prime space P_f with elements (p_x, f_g). In such a case, we normally assume the standard depth of the space to be 10^5. Now we noted that the gaps "2," "4," and "6" define a base for other gaps. That is,

a. it was first noted in the frequency curve which had repeated behaviors along the curve;
b. then it was observed that the equation $p_1 = 6a \pm 1$ did not have a universal characteristic. It has a particular application described by the gap base and consequent gaps structured along the gap base. That is $p_1 = 6a + 1$ was for gap "2," and $p_1 = 6a - 1$ was for gap "4" while gap "6" had the format $p_1 = 6a \pm 1$. This also helped in classifying the prime numbers using a gap theory.

Now if all the primes of a given family (p_x, f_g) are plotted on a graph, the surprising results that we get a linear relationship. This is based on the description $f(t, p_t) = m_x t + k$.

In this relationship, we note that,

a. the set of primes $(p_x, 2)$ are almost parallel to $(p_x, 4)$, the gradient is about equal, and
b. the set of primes $(p_x, 6)$ are at a smaller gradient with respect to the other two families.

This demonstrates that prime numbers behave as a family in terms of their distribution. That is, they do not normally follow the sequential distribution pattern that is expected of them. This pattern of behavior is defined as a spectral distribution of primes, and the graph is called spectral lines of prime numbers.

THE THEORY OF PRIME NUMBER CLASSIFICATION

The concept of the gap base defines a class of primes. The next three groups give the next class of prime gaps and so on. Now as the class number goes up, the straight line aspect of the distribution of the primes along the line is lost; they become more rugged. However, the relationship that exists in terms of the base structure of the gaps is still very strong.

By using a best line fit, each of the spectral lines of the primes has a gradient as shown on the graph and table below. The gradient m of the spectral lines are defined to be directly proportional to the number of primes, hence giving us a measure of their density. Typically density is defined by a count function.

From this observation, prime density $P_d(x)$ is defined as the inverse of the gradient multiplied for the natural gap by 1000. Later, a probabilistic density function called the P-index is also defined.

Definition 2

A prime family (p_x, f_g) in dimension x has prime density $P_d(x) = 1000/m_f$, where m_f is the gradient for a given f_g.

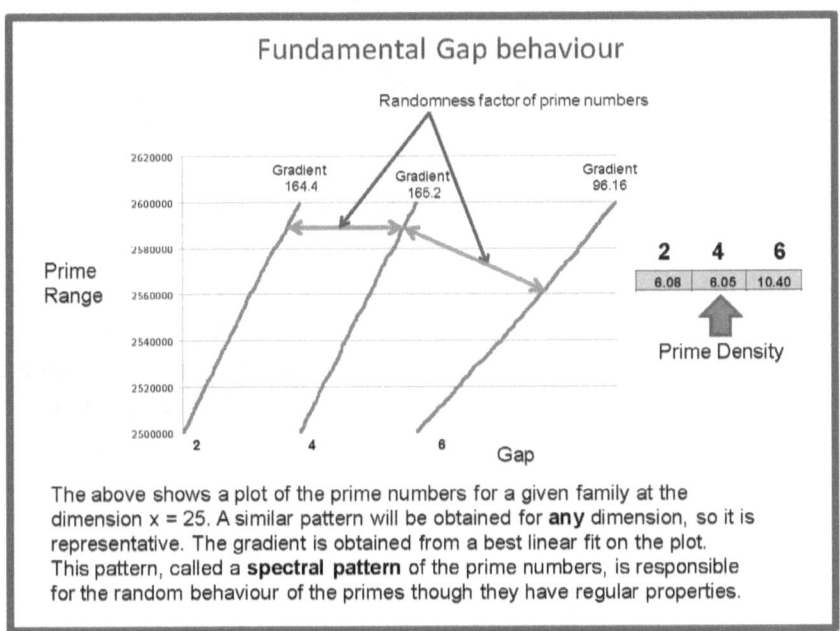

The above shows a plot of the prime numbers for a given family at the dimension x = 25. A similar pattern will be obtained for **any** dimension, so it is representative. The gradient is obtained from a best linear fit on the plot. This pattern, called a **spectral pattern** of the prime numbers, is responsible for the random behaviour of the primes though they have regular properties.

Figure 38. The gradient factor in prime gap relationships.

The density gives us a relative measure through which we can use gaps to study how prime numbers are distributed.

	Gap 2	Gap 4	Gap 6	Gap 44	Gap 46	Gap 48
Gradient	164.4	155.2	96.16	3928	3985	2839
Prime Density	6.08	6.05	10.40	0.2546	0.2509	0.3522

Table 22. Comparing density in the same dimension for given prime families.

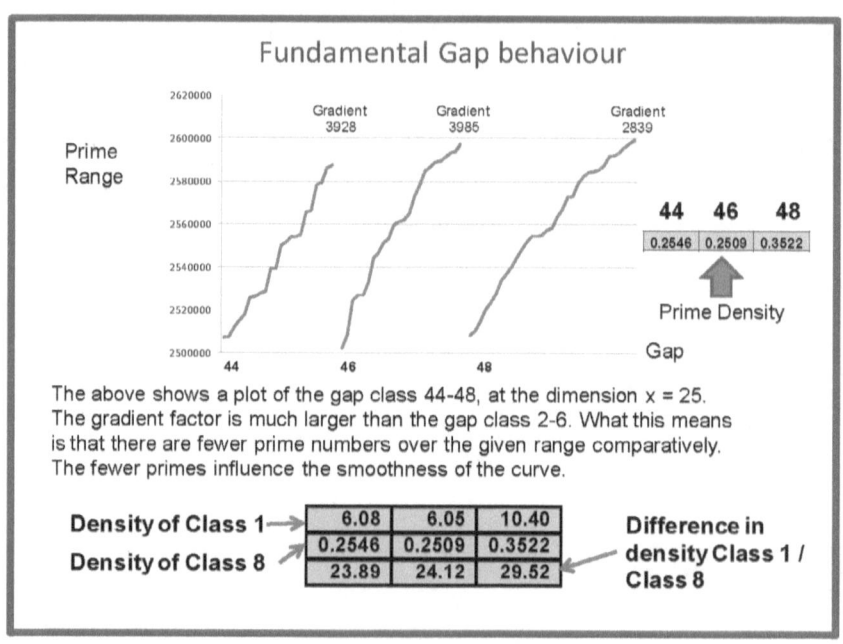

Figure 39. Prime density for large gaps.

For example, the actual prime count for gap 2, 4, 6 is 1227, 1216, and 1941 respectively, and this is reflected in the ratio of prime density. We then note that as the prime class increases, the prime density decreases, where this decrease also seems to be proportional. For example, if we compare class 2-4-6 to class 44-46-48, then we have a proportional decrease of 23.89, 24.12, 29.52 respectively.

THE THEORY OF PRIME NUMBER CLASSIFICATION

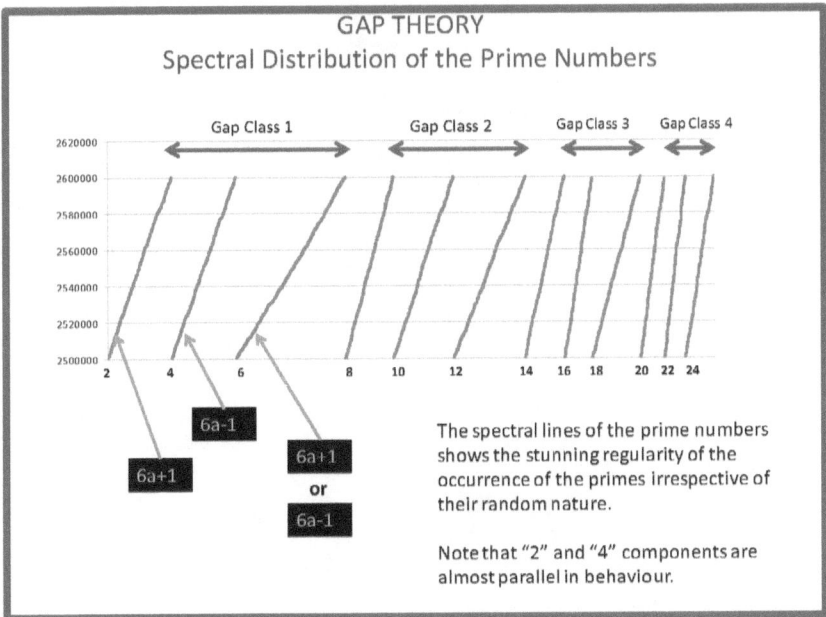

Figure 40. Showing gap class and spectral lines and prime equations.

The above graphics also point out the fact that we should not just view primes in a sequential manner. Yes indeed, prime numbers are in a sequence, the next prime is greater than the previous prime, but they are not in the same group or category because of that. Therefore, the aspect of being sequential is not an overriding factor in understanding their distribution behavior. The graphics demonstrate a parallel behavior in terms of the distributive property, as well as in terms of the prime density.

4 FUNDAMENTAL GAP BEHAVIOR—SPECTRAL LINES

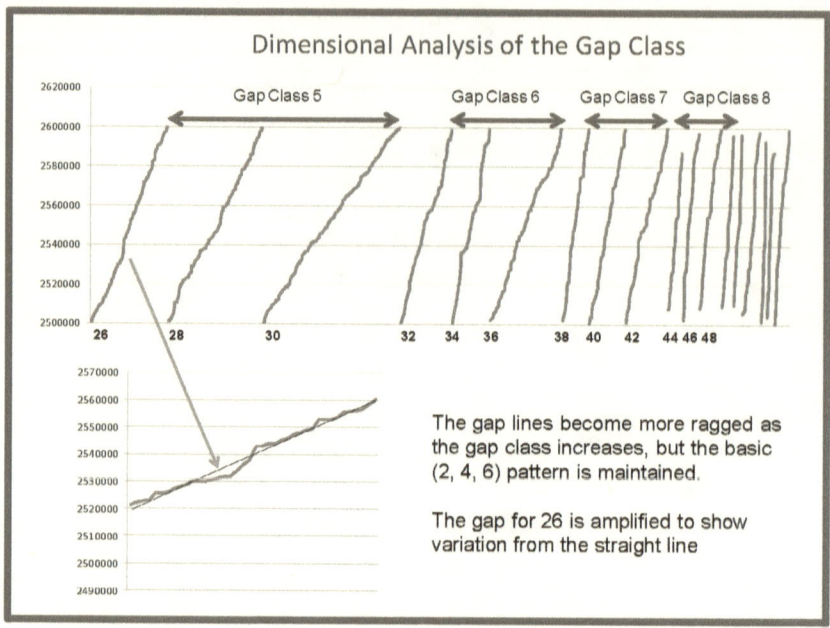

Figure 41. Spectral lines become more ragged as gap increases.

This observation is important if we have to consider the concept of prime number generation. It implies that we can base prime number generation on the gap theory because distribution, gaps, and generation are all related to the final prime number and where it is in the number spectrum.

Section Five

Gap Theory of Prime Number Generation

Overview

The equation $p = 6a \pm 1$ is used as an analytical tool to further study the random nature and generation patterns of the prime numbers.

Objective:

To study the predictive and linear behavior of primes numbers in a two-dimensional space.

Research Question

a. How does prime number classification assist in understanding prime number generation?
b. Is random behavior of prime numbers defined as a consequence of a sequential process only, or can it be nonsequential and be based on underlying sets relating to form a larger whole?

1 Introduction

One of the key research questions is, how does classification of prime numbers assist in understanding prime number generation? One of the techniques of generating primes over the centuries has been the Sieve of Eratosthenes, and consequently, classification and prime number generation have never been directly linked. In a similar manner, classification has not been seen as key to understanding the random behavior of the prime number. The advance of computing has also made prime number generation an arithmetic issue rather than a theoretical issue.

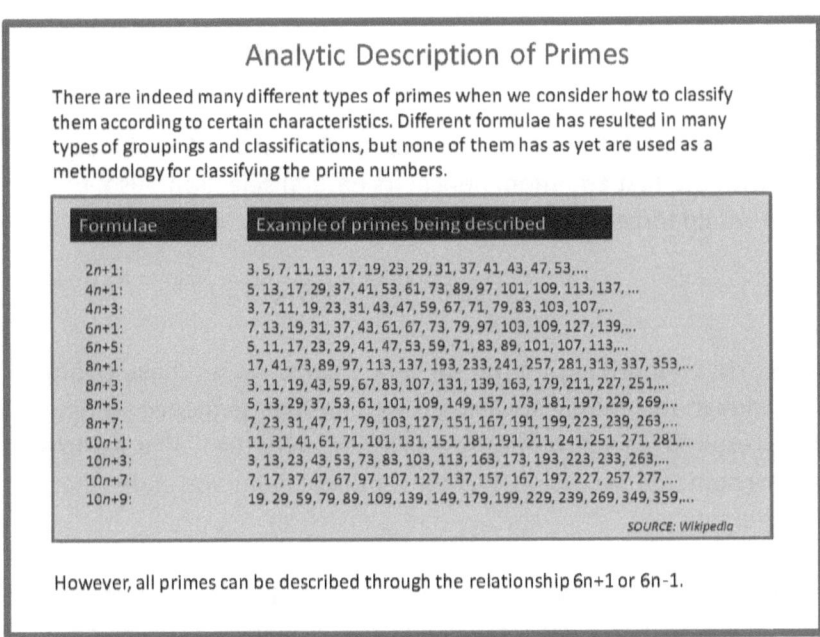

Figure 42. The different formulas for prime numbers.

1 INTRODUCTION

It is uncommon to classify primes using an arithmetic expression or formula because one formula is able to produce only certain primes while another produces a different set as shown in the above diagram. Some prime numbers may even appear as common between two or more such formulaic descriptions. Also, the common dominant approach has been to find one formulaic description for all primes, whereas the reality is that this may not be the case.

In the context of this theory, there are two fundamental conditions that define prime numbers in the natural set of numbers:

1. Any natural number can be written in the form $6a \pm b$, where a is any integer, and $b \in H = \{0, 1, 2, 3, 4, 5\}$
2. Any prime p number satisfies $p \in \{2, 3, 6a \pm 1\}$ and $a \in N$, the set of natural numbers.

Taking these two conditions mean there are actually two formulas that we may use to generate prime numbers, and these are the following:

$$p = 6a \pm 1$$

$$p = 6a \pm 5$$

Using either of the formulas will give us similar result. It is obviously easier to work with the first equation, hence, its natural selection. Also, the second equation would force us to redefine as follows:

$$p \in \{2, 3, 5, 7, 6a \pm 5\}s$$

It allows 1 to be a prime number for $a = 1$ since we will have two results 1 and 11, though technically speaking if we consider sequence within the set, then the result "1" is automatically disallowed. Whether "1" is a prime or not is always a point for discussion, but conventional definition states that it is not a prime number.

The other common observation is the role of $6a$. This has been shown in the previous chapter that it has influence in the formation of prime gaps. There has been previously no link to the fact that $6a$ has influence in the formation of prime gaps. Therefore, it is this observation that has actually linked the formula $p = 6a \pm 1$ to form the prime gaps theory. In the previous chapter, we considered mostly gap behavior in respect to classification, but now we would like to focus on prime number generation with respect to classification. Part of this also involves looking into the random nature of the prime number since this goes hand in hand with prime number generation.

2 The Random Function of Prime Numbers

Consider the following numbers produced by the given functions for $x = 1, 2, 3, 4$. Then,

A = {5, 7, 9, 11), where these are elements of the set $y = 2x + 3$.
B = {2, 7, 12, 17}, where these are elements of $y = 5x - 3$.
C = {1, 4, 9, 16}, where these are elements of the set $y = x^2$.

Now if we consider a union of the above sets, where we place the condition of taking the first four elements only to form such a set U, then we have,

$$U = \{1, 2, 4, 5, 7, 9, 11, 12, 13, 16, 17\}$$

The approximation equation that describes the elements of set U will be $y = 1.645x - 1.054$. If one had begun with this equation, it would indeed be very difficult to work back to the other three equations unless we used pattern recognition between the elements. The values of the equation can represented by set U_2, where this would give us U if we rounded.

U_2 = {0.591, 2.236, 3.881, 5.526, 7.171, 8.816, 10.461, 12.106, 13.751, 15.396, 17.041}

Now we can call the first A, B, and C as underlying sets that defined the exact nature of the elements or condition we seek to describe, then set U becomes the observable set, and U_2 becomes the approximating set which we use to represent U for purposes of having a mathematical statement about U. This is the type of challenge that is presented by the prime numbers. We can derive the set of prime numbers, and they form the observable set P. Because of the random nature in P, we then use or derive approximating equations in order to

THE THEORY OF PRIME NUMBER CLASSIFICATION

understand their behavior. One of the alternative approaches is to determine if there are underlying sets that can assist us to get back to the original description of the elements, hence the concept of the prime family.

We can also make the following observations about the sets A, B, and C:

a. The elements of each set can be described by a dependent variable.
b. The elements of each set are predictable.
c. The gap between the elements is consistent with the equation describing the elements.

However, the set U does not consist of predictable elements. This we note from the gap between the elements.

	1	2	4	5	7	9	11	12	13	16	17
Gap		1	2	1	2	2	2	1	1	3	1

Table 23. Showing the random nature because of underlying sets.

From this, we then consider the set U to consist of random elements. Random behavior is recognized by the inability of a pattern to demonstrate a systematic repetition. On the other hand, predictability has to do with the ability to be certain about a common systematic pattern to the given numbers. Hence, the concept of randomness is linked to the ability to predict values. What is common between the two are as follows:

a. *Distribution*: does the distribution pattern fit some given formula or show signs of consistency versus inconsistency?
b. *Sequencing*: is there a particular order in the distribution, and can the sequencing allow some relation to be drawn again by looking at consistency versus inconsistency?

This is the primary framework around which the prime numbers have been studied and continue to be studied. Now while prime numbers have been observed to be random in occurrence, they also demonstrate predictable behavior patterns. This then is an indication of some underlying pattern, and this is what has given hope to mathematicians to keep on looking for an algorithm that will describe such an occurrence. We therefore consider the following fundamental question:

2 THE RANDOM FUNCTION OF PRIME NUMBERS

Is random behavior of prime numbers defined as a consequence of a sequential process only, or can it be nonsequential and be based on underlying sets relating to form a larger whole?

That is, the following is conjectured:

Conjecture 1

Prime number incidence is non-sequential

The logic here is that for the universal set of prime numbers, indeed, we observe a sequential behavior, and the argument is that we observe a sequence because we are studying them on a one-dimensional basis. That is, the incidence of a prime number occurs at the underlying set, not at the universal set. Therefore, instead of using one formula to establish the patterns for deriving or locating a prime number, it is better to consider different underlying sets of primes coming together to form a sequence that is random.

To reconcile the sequential behavior at both underlying sets and the universal set of primes, we must then assume the following two propositions:

Proposition 1

Let P_u be the universal set of prime numbers consisiting of prime family subsets $f_1, f_2, \ldots f_n$, where $f_1 \cap f_2 \cap \ldots \cap f_n = \emptyset$

This proposition has the consequence that whatever element we have for the underlying sets, because of the uniqueness of the elements of the subsets, then these will result in a sequence of numbers in P_u. Secondly,

Proposition 2

There exists a randomizing function of prime numbers $RND(p)$.

The assumption is that such a function does not operate at the universal level P_u, but operates mainly at the level of the prime number subsets, that is f_n. Now we know the description of each f_n because of the gap base theory. That is, the

THE THEORY OF PRIME NUMBER CLASSIFICATION

gap base 2 is defined by $p = 6a + 1$, the gap base 4 is defined by $p = 6a - 1$, and the gap base 6 by both of them. Therefore, we assume as follows:

Definition 1

From the fact that for any prime sequence $q_1 = (b - a)$, then let the random function be defined as $RND(p) = f(a)$

That is, for prime p_1 and p_2, let $p_1(a) = a_1$ and $p_2(a) = a_2$ and $a_2 > a_1$ since the primes are in sequence also within the family, then let,

a. $f(a) = a_2 - a_1$. If $f(a) = k$, a constant, then $RND(p)$ is not a random function, and the prime numbers will be exactly predictable.
b. $f(a) = a_2 - a_1$. If $f(a) \neq k$, a constant, then $RND(p)$ is a random function, and the prime numbers will be not predictable.

Therefore, for gap base 2, where $p_1 = 6a_1 + 1$, solving gives,

$$a_1 = (p_1 - 1)/6$$

$$a_2 = (p_2 - 1)/6$$

Hence, the random function for prime numbers in this family is defined by,

$$a_1 - a_2 = (p_2 - p_1)/6$$

The same equation is derived for gap base 4. For gap base 6, the above equation applies including the following options that were referred to as crossing in the classification.

$$a_1 = (p_1 + 1)/6$$

$$a_2 = (p_2 - 1)/6$$

This derives the random function,

$$a_1 - a_2 = (p_2 - p_1 - 2)/6$$

2 THE RANDOM FUNCTION OF PRIME NUMBERS

Similarly for,

$$a_1 = (p_1 - 1)/6$$

$$a_2 = (p_2 + 1)/6$$

This gives the random function,

$$a_1 - a_2 = (p_2 - p_1 + 2)/6$$

The graphic below shows the first few values for $a_1 - a_2$ for gap 2.

Family	Prime	Minus 1	Value of "a"	RND value for Gap 2
2	5	4	0.66666667	
2	7	6	1	
2	13	12	2	1
2	19	18	3	1
2	31	30	5	2
2	43	42	7	2
2	61	60	10	3
2	73	72	12	2
2	103	102	17	5
2	109	108	18	1
2	139	138	23	5
2	151	150	25	2
2	181	180	30	5
2	193	192	32	2
2	199	198	33	1
2	229	228	38	5
2	241	240	40	2

Figure 43. Calculating the random function for gap 2.

THE THEORY OF PRIME NUMBER CLASSIFICATION

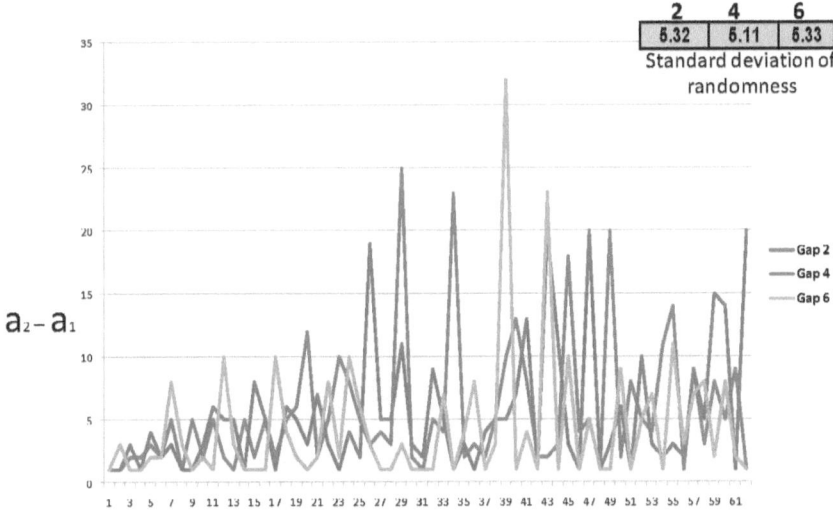

Figure 44. Graph of the random function.

It is of interest to find out what is the random function generator for the prime numbers. The relationship is known, but what is it that causes or defines the random function? This is a question of interest since it has been noted that prime numbers have a linear relationship, so something must be disturbing them to make them to have a random structure.

Hence, we see that classification and the randomness of prime numbers have been connected through these equations. It makes sense because really, classification is the study of the resultant behavior of the prime generation as a consequence of the equations.

If we take the average values $RND(p)$ for a given gap base over a given similar range, then the average is highest for gap base 2, next is gap base 4, and lastly gap base 6.

2 THE RANDOM FUNCTION OF PRIME NUMBERS

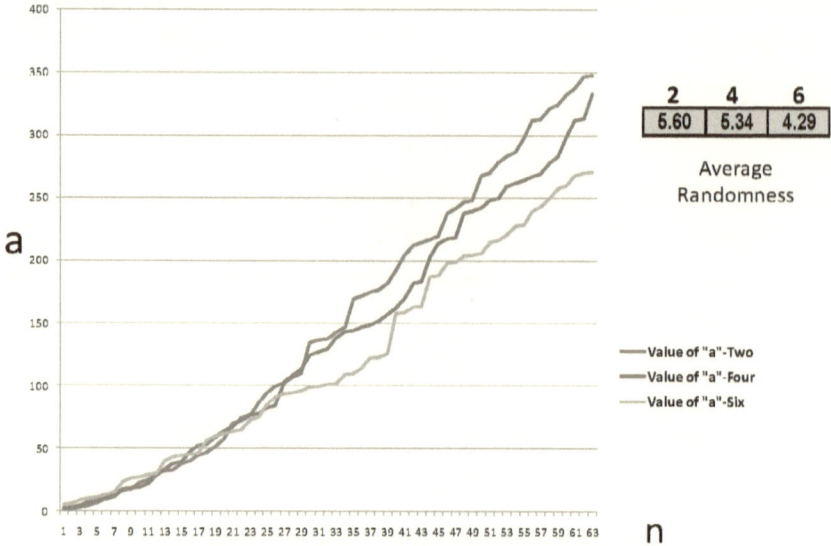

Figure 45. Graph of the values of "a" for each gap.

Since a random function has been defined, the average randomness of a function can also be determined for a given range for the same dimension.

	Gap Base 2	Gap Base 4	Gap Base 6
Standard deviation of random behavior	5.32	5.11	5.33
Average value of randomness	5.6	5.34	4.29

Table 24. The comparison of randomness between gaps.

One would have expected that gap base 6 would be more random given that it is described by more than one formula to derive a prime number. It is surprising that gap base 2 primes are leading in random behavior. The research questions are these:

> *Is there a specific reason for smaller gaps being more random in behavior?*
>
> *Is this a consistent behavior even at infinity?*

Perhaps a larger range will show a different result since this was taken for the first sixty primes or so. We expect a similar pattern for gaps 8-10-12, that gap

THE THEORY OF PRIME NUMBER CLASSIFICATION

8 will show higher randomness than 10 and 12 because of the gap base theory. However, the standard deviation of the random behavior is almost the same, indicating that the behavior pattern within the gap bases is similar.

The usefulness of the above analysis is that when we look at a gap base, we may consider three main attributes to describe it. These are the following:

1. The density of primes numbers $P_d(x)$ for the given range in the gap base.
2. The nature of the random behavior in the given range and gap base through standard deviation.
3. The general trend of the randomness of the primes in the gap base for the given range through average values of "a."

The density of the primes was defined in terms of the gradient of a particular range in a given family, while randomness is determined in relation to "a."

3 The Generation of Prime Numbers

Finally, we now consider the generation of prime numbers in the context of the classification theory. We go back to the same basic expression that all primes can generated from $p = 6a \pm 1$. The difference is the fact that we now apply the gap theory when we consider the generation of the prime numbers. This has the implication that is shown in the diagram of figure 45.

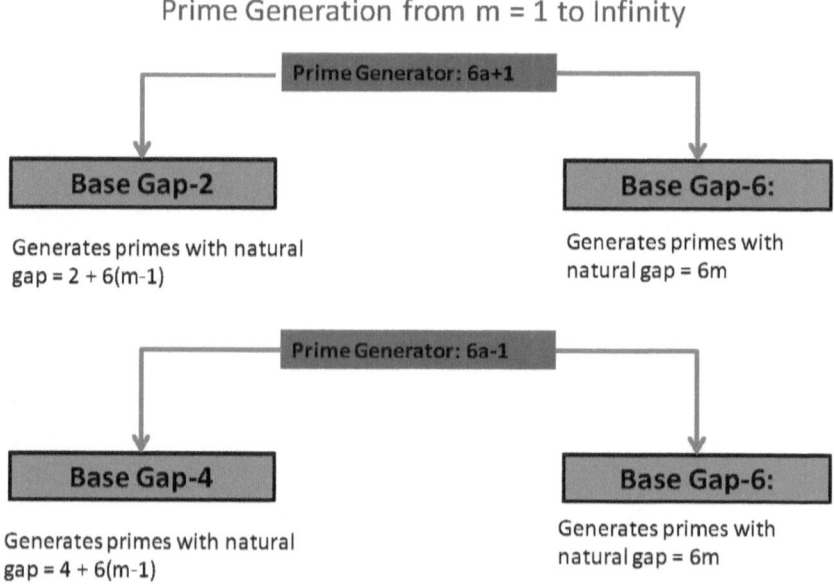

Figure 46. The description of the prime number generators.

Fundamental Gap Theorem

The set of all prime numbers is generated by $p = 6a \pm 1$, where,

THE THEORY OF PRIME NUMBER CLASSIFICATION

a. the expression $p = 6a + 1$ generates all prime numbers in gap base 2, that is, those with the natural gap $g(m) = 2 + 6(m - 1)$, and those with natural gap $g(m) = 6m, 1 \leq m \leq \infty$;

b. the expression $p = 6a - 1$ generates all the primes with gap base 4, that is, those with the natural gap $g(m) = 4 + 6(m - 1)$, and those with natural gap $g(m) = 6m, 1 \leq m \leq \infty$.

c. In a given gap base, the gap between any two prime numbers is a multiple of six.

The theorem leads to the following consequence.

Definition 2

In prime number generation, gap base 6 is a subset of gap base 2 or gap base 4.

In other words, only two gap bases are involved in prime number generation: gap base 2 and gap base 4. The other consequence of the fundamental gap theorem is that it also generates numbers that are not prime but are referred to as pseudoprimes. The reference of a pseudoprime as a composite is a misnomer in this theory.

Definition 3

For the prime generator $p = 6a \pm 1$, if p is not prime, then it is a pseudoprime.

The generator has two outcomes, the prime number and the pseudoprimes, and the same rules about gaps applies to both. Normally, a prime number generator is defined to only generate prime numbers. The reason for this approach will be evident in later application of the generator. For example, the Sieve of Eratosthenes is a prime number-generating construct. The difference is that this one is based on the gap theory, and secondly, it also includes numbers that are not prime.

We can now work toward a particular definition of a prime space that allows us to generalize the prime generation context. Consider the following, which summarizes some of the previous definitions.

3 THE GENERATION OF PRIME NUMBERS

a. $P_u = P_s$, where $P_s = [1, 1, \infty]$, where this is the conventional prime number space
b. $P_u = P_s$, where $P_s = [1, d, \infty]$, where $d \geq 10^5$, where this is a two dimensional space
c. Let $f_n \in P_f$, where $1 \leq n \leq \infty$

The last point is of interest, that is, whether there are infinitely many prime families. For the purposes of this theory, it is assumed that this is the case. The value of d has been determined as being reasonably large for a small study at $d = 10^5$. This does not mean it cannot be smaller, it is chosen as being conveniently small. We have a universal prime space and a two-dimensional space, and f_n is a subset of the universal space and defines a prime family in the form (p_x, f_g), where x is a dimension and f_g the gap between primes of the same family. Therefore,

$$f_n = (p_x, f_g = 2n)$$

For example, $f_1 = (p_x, 2)$ describes all the primes with gap 2 in dimension x. Similarly, $f_2 = (p_x, 4)$ defines the family with gap 4 but still in the dimension x. Now f_n will be defined differently, that is, for a given n, the dimension x varies from one to infinity while m, which defines the natural gap, also varies likewise. This is because the framework is that of prime generation, so we need values that change to define the next prime. In the first definition, the dimension was held constant. This is the basic difference between the space P_u and H_p that is going to be defined. Both consist of prime families f_n, but in one, the dimension is held constant, while in the other the space is also divided into three distinct groups that define the gap base for the prime space. Note that x defines dimension.

Prime Generation Axioms

1. There exists an initial prime $p_0 = 6a_0 \pm 1$, defined by a_0

2. *Prime generation condition*: Let $a_n = a_0 + n$, and $1 \leq n \leq \infty$, n an integer.

3. There exist a prime number or pseudoprime $p_n = 6a_n \pm 1$

4. Let there be a prime family space with elements p, where

THE THEORY OF PRIME NUMBER CLASSIFICATION

a. $f_n(1) = (p, f_g = 2 + 6(m-1), x)$, $1 \leq x \leq \infty$, $1 \leq m \leq \infty$ for a given n.

b. $f_n(2) = (p, f_g = 4 + 6(m-1), x)$, $1 \leq x \leq \infty$, $1 \leq m \leq \infty$ for a given n.

c. $f_n(3) = (p, f_g = 6m, x)$, $1 \leq x \leq \infty$, $1 \leq m \leq \infty$ for a given n.

5. Let there be a prime space $H_p = f_n(1) \cup f_n(2) \cup f_n(3)$ called the linear delta space.

6. Let $p \in f_n(3)$, then $p \in f_n(1)$ or $p \in f_n(2)$

7. $f_n(1) \cap f_n(2) = \emptyset$

8. $a_1 - a_2 = \frac{(p_2 - p_1 \pm R)}{6}$, where $R = 0, 2$

9. Let there a natural number $q \in H_s$ the pseudoprime space, then $q \notin H_p$, $p \in H_p$, and $p - q = 6h$ within the same gap base.

The above axioms assume an initial prime because we are considering prime generation must start at a given point, and the third axiom defines the generation process. The fourth axiom defines the partitions of the space H_p, which is also a universal space. Axiom 5 formally defines the linear delta space as the prime generation space. Axiom 6 states that basically there will be two resultant sets from the prime generation (a consequence of the fundamental gap theorem), and axiom 7 says these sets are mutually exclusive. Axiom 8 then states that the randomizing function must apply to all elements of the defined sets. Axiom 9 acknowledges the fact that the equation for prime generation also produces odd numbers in set H_s that are not prime, and because they have the same gap relation to the real prime number, they are referred to as pseudoprimes. It is probable that this axiom could have been left out, but the inclusion of pseudoprimes as part of the generation process is actually critical for the demonstration of the linear properties of prime numbers and establishing behavioral properties of the primes.

Now in terms of prime generation, $p_n = 6a_n \pm 1$, and if we substitute $a_n = a_0 + n$, this expands to give us the gap-base equation for generation. That is, $p_n = 6a_n \pm 1$ is a classification equation, but the one below is a gap generation equation, it builds the set in terms of a given consistent gap $6n$.

3 THE GENERATION OF PRIME NUMBERS

$$p_n = 6a_0 + 6n \pm 1 = p_0 + 6n, 1 \leq n \leq \infty$$

That is, all prime numbers in the respective prime families will have a gap that is a multiple of six, inclusive of pseudoprimes. Since there are only two sets to consider and $f_n(1) \cap f_n(2) = \emptyset$, then we have to identify the intial prime number p_0 for generation that satisfies each of the sets.

Now these possible values of a_0 are defined by $p_0 = 6a_0 \pm 1$, where such a choice should not be a decimal.

	$p_0 = 6a_0 + 1$	$p_0 = 6a_0 - 1$
	Gap Base 2	Gap Base 4
$p_0 = 1$	$a_0 = 0$	$a_0 = 2/6$
$p_0 = 5$	$a_0 = 4/6$	$a_0 = 1$
$p_0 = 7$	$a_0 = 1$	$a_0 = 8/6$
$p_0 = 11$	$a_0 = 10/6$	$a_0 = 2$
$p_0 = 13$	$a_0 = 2$	$a_0 = 14/6$

Table 25. Determining the initial values.

We assume that if we have a_0 for gap base 2, then for gap base 4 it must have $a_0 + 1$ for the next prime occurrence. Therefore, from the table, possible values are $a_0 = 0$ and $a_0 = 1$, or $a_0 = 1$ and $a_0 = 2$ for gap base 2 and 4 respectively.

Definition 3

Let $a_0 = 1$ for gap base 2 and $a_0 = 2$ for gap base 4.

There is no justification for 1 to be a prime number as yet, hence the pair $a_0 = 0$ and $a_0 = 1$ are ignored. The next logical choice then is $a_0 = 1$ and $a_0 = 2$. Therefore, substituting in $p_n = p_0 + 6n$ gives us the following prime number-generating equations.

1. $p_n = 7 + 6n$ for $p = 6a + 1$, that is gap base 2 in $f_n(1)$ and gap base 6 primes in $f_n(3)$

THE THEORY OF PRIME NUMBER CLASSIFICATION

2. $p_n = 11 + 6n$ for $p = 6a - 1$, that is gap base 4 in $f_n(2)$ and gap base 6 primes in $f_n(3)$

These form the basic framework for the prime and pseudoprime generating function.

4 The Multi-value Function for Prime Numbers

Generally, $y = f(x)$ is the standard approach in the concept of functions and defines a one-to-one mapping. If we want to extend the dimensionality of the function, then we consider $y = f(x_1, x_2, \ldots x_n)$ also as a one-to-one mapping as it represents one point in a multidimensional space. However, if we consider a circle whose general equation is $x^2 + y^2 = r^2$, then there are two possible results for y, that is, $\pm y$ for the every x—a one-to-many mapping. The example of the circular function shows a multivalued function in the same dimension, that is, having more than one point to represent the function but still being in the same dimension.

Whilst we note that we can actually have a one-to-many mapping within the same dimension, consideration is not normally made that the same can be true for linear functions. The development of uniqueness in regard to the one-to-one mapping of the function has meant that this technique may have been seen as not representing real-life situations to problems or mathematical context. It is for this reason that most attempts to describe the prime numbers in formulaic expressions have tended to look for one equation with one-to-one properties for prediction. The relationship between prime numbers provides evidence to conceptualize a linear space that is multi-valued and presents this as a problem-solving technique.

> **Definition 4**
>
> Le $f(x) = (y_1, y_2, \ldots y_n)$ be an n-dimensional multi-valued function provided $(y_1 \cap y_2 \cap \ldots \cap y_n) = \emptyset$.

The condition defines a uniqueness for the function, that is, it is a one-to-many, but the union of all the elements of the function will be unique within the

THE THEORY OF PRIME NUMBER CLASSIFICATION

set. It is not a simultaneous equation because of the way the solution for the equations is defined because, in such an equation, we would have $y_1 = f(x_1)$ and $y_2 = f(x_2)$. However, $n = 1$ gives us the function $y = f(x)$ as a one-dimensional multi-valued function. The function $y = f(x)$ is called a normal or standard function while for $n > 1$, we then have a random relation function.

Note that the context of definition of the function is not with respect to a geometric relationship, it is in the context of an algebraic space. Therefore, $f(x) = (x^2 + 3, x)$ is an example of a mulitvalued function, but $f(x) = (x^2, 4x + 5)$ is not because of intersection. Now what we have done is to provide a general framework of understanding for a particular type of function. We then apply this in the context of the prime number-generation assumptions.

Consider the two equations $p_n = 7 + 6n$ and $p_n = 11 + 6n$, then it was determined that $p_0 = 7$ and $p_0 = 11$ for gap base 2 and 4 respectively. This then determines our value for x = 0 on the y-axis. Hence, from this, knowing that every prime or pseudoprime has a gap of 6 in between for the same gap base, we then derive two equations as follows:

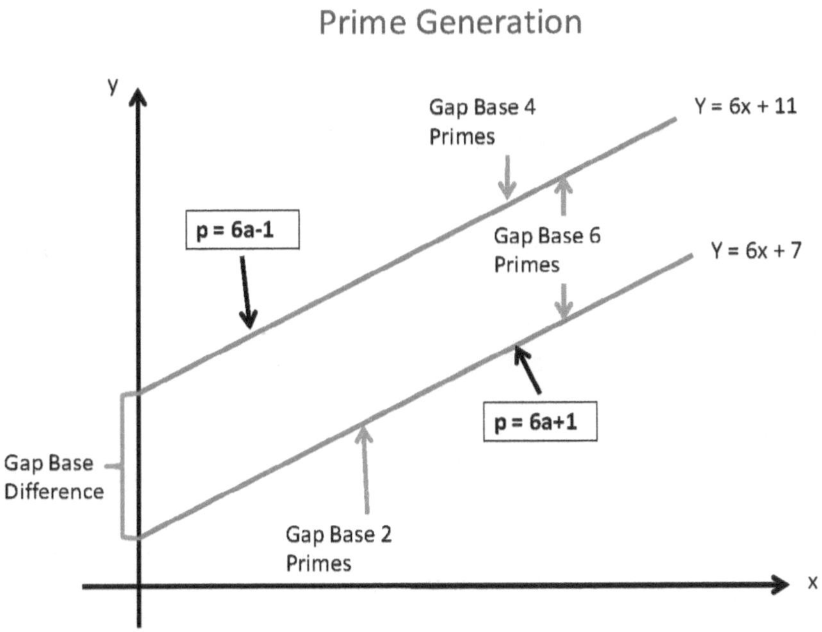

Prime Generation

Figure 47. The concept of prime number generation.

4 THE MULTI-VALUE FUNCTION FOR PRIME NUMBERS

$$y_1 = 6x + 7, where\ 0 \leq \text{Int}\ x \leq \infty$$

That is for gap base 2 and gap base 6. Similarly

$$y_2 = 6x + 11, where\ 0 \leq \text{Int}\ x \leq \infty$$

That is for gap base 4 and gap base 6 starting at $x = 0$. Later, as the theory develops, we shall start at $x = -1$.

The use of this type of function in describing prime numbers has two consequences:

a. This results in a strict description of the pseudoprimes as being related structurally to primes.
b. It provides an opportunity to develop a sieve structure for deriving primes according to gap and classification.

It also causes a shift from considering a solution based on a one-to-one relationship. The two formulas allow us to relate the equation $p = 6a \pm 1$ to describe a linear relationship between the primes and, in so doing, also providing a simple sieve that conforms to a classification theory based on gaps.

Color coding is again used to demonstrate the complexity of relationships in the prime-generating theory. Values are taken for $x = 1$ to 17, and generation takes place.

1. The yellow code shows primes that belong to gap base 2 only. It also shows primes that have gap two, that is, belong to the family $(p, 2)$.
2. At x = 15, then we have the first occurrence for a prime for the family $(p, 8) = (97, 8)$, that is, with gap 8. The next occurrence is at x = 60, where we have $(p, 8) = (367, 8)$, and so on.
3. The orange shows primes that belong to gap base 4 only. The first occurrence of $(p, 10) = (149, 10)$ at x = 23. It does not show since the analysis ends at x = 17.
4. The olive green shows elements of gap base 6 that are generated either in gap base 2 or gap base 4. Note that we do not have a_0 for gap base 6 because it is defined in the context of gap base 2 or 4.

5. The light purple gives us the set of pseudoprimes, that is, the set $q \in H_s$. Since H_p consists of prime numbers only; therefore, it implies $q \notin H_p$.

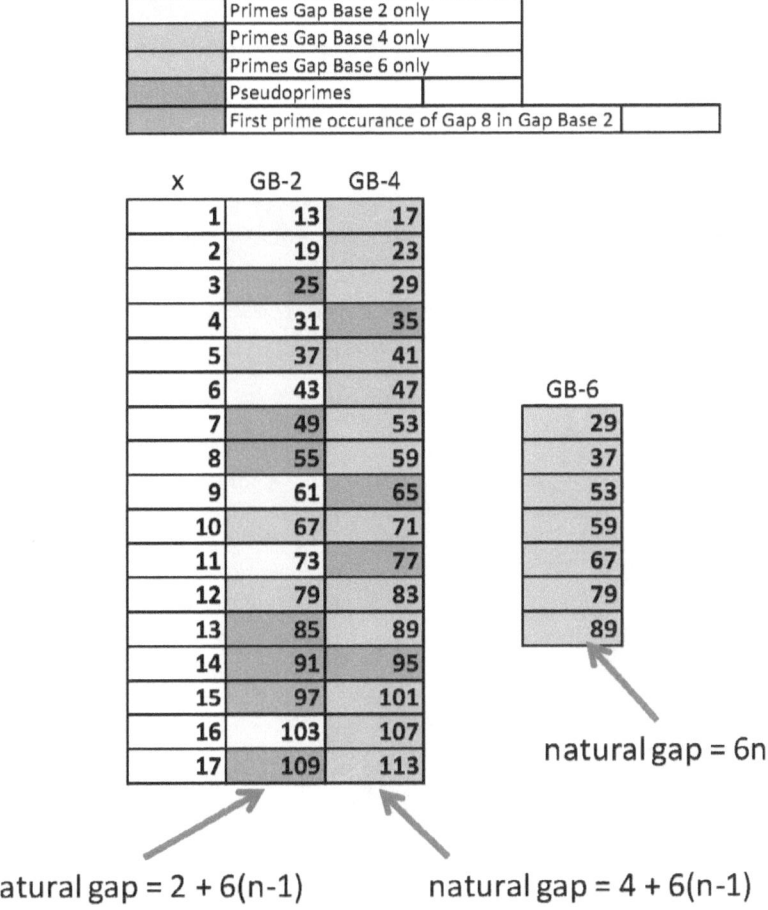

Figure 48. Showing generation according to the gap theory

1. What is key to observe with this generation theory is that we use the same value of x. For example, $x = 15$ derives the primes $(97, 8)$ and $(101, 4)$.

For same value of x	Gap Base 2 y_1	Gap Base 4 y_2
y	Prime	Prime

4 THE MULTI-VALUE FUNCTION FOR PRIME NUMBERS

y	Prime	Pseudoprime
y	Pseudoprime	Prime
y	Pseudoprime	Pseudoprime

Table 26. Possible outcomes for generation.

We also observe that the sequence of the prime numbers is still maintained. That is, using gap theory does not imply that some prime numbers will be skipped. Lastly, we observe that by selecting specific values of a_0, this also defines specific values of pseudoprimes. For each x, there is a unique value of a pseudoprime, and the pseudoprimes are also random in occurrence like the prime numbers. Because of this relationship, these are referred to as natural pseudoprimes.

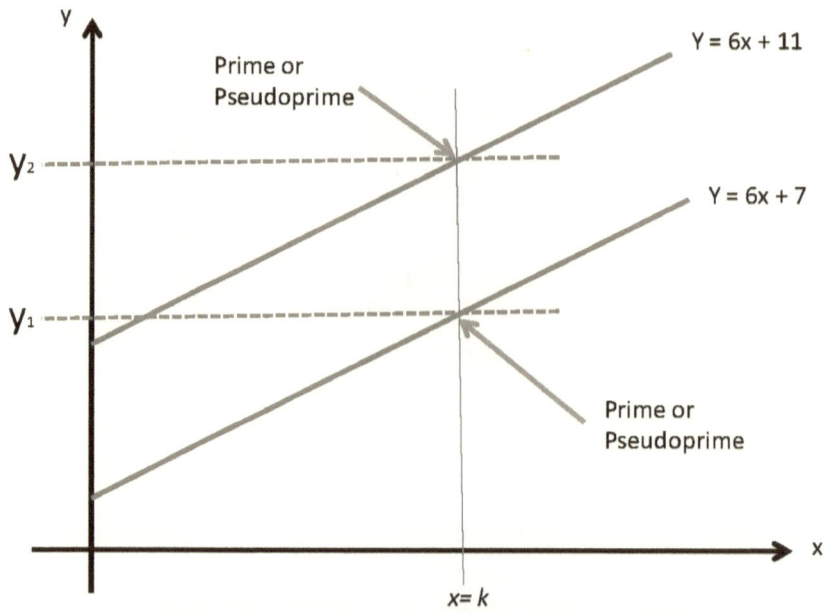

Figure 49. Graphical relationship of outcomes.

Therefore, in the above example, as we generate the prime numbers, we also generate unique pseudoprimes as follows for $x = 1$ to $x = 17$ taken from both gap bases.

$$H_s(1,17) = \{25, 35, 49, 55, 65, 77, 85, 91, 95, 109\}$$

They form a specific sequence of numbers in the same way as the prime numbers, and the gap between the pseudoprimes is also random.

$$\text{Pseudoprime Gaps }(1, 17) = \{10, 14, 6, 10, 12, 8, 6, 4\}$$

However, note that the relationship between primes and pseudoprimes expressed as $p - q = 6h$ also holds. For example, take $x = 7$ with pseudoprime 49, and $x = 11$, which is a prime 73, then (73-49) = 24; hence, $h = 4$.

It is normally neater in mathematics to start with a theorem and prove it true. On the contrary, because I believe expository mathematics creates a knowledge scenario that allows theorems to be born, I will conclude this research with the following theorem.

Theorem

All prime numbers exist either on $y = 6x + 7$, or $y = 6x + 11$, where $-\infty \leq Int\, x \leq \infty$.

This theorem is stated in contrast to the already accepted mathematical formulas that are said to partially generate the prime numbers until some given value of n as indicated in the introduction of this section. The equations do not describe a formula for finding the nth value of a prime number for a given n, but merely affirm where we expect to find prime numbers algebraically. Therefore it has the consequence of narrowing the search for them, hence the assumption of the existence of a primality test based on the theorem.

It may be noted that the equations $y = 6x + 1$ and $y = 6x + 5$ are equally true for the theorem if values of a_0 are 0 and 1 respectively. Later the generality of choice will be established.

Research Conclusion

The fact that prime numbers lie on either lines explains their regularity as well as random nature (Sunday, June 27, 2010).

SECTION SIX

THE ALGEBRAIC SIEVE

Overview

The equation $p = 6a \pm 1$ is used as an analytical tool to further study the random nature and generation patterns of the prime numbers.

Objective:

To develop an arithmetic sieve that predicts the nth prime number. (This objective did not succeed, but it has been left to show the dynamics of expository mathematics, and that something can still be attained as one plods along the way of discovery).

Research Question

How can the sieve provide a framework for primality testing?

Is it possible to use the sieve to prove the twin-prime conjecture?

1 Introduction

The next area of interest after one considers the generation of the prime numbers is the prediction of the nth prime number. Sieve theory has been used as a tool to generate prime numbers, but currently, they do not have predictive power, and their major focus is generation. However, the aim is now to develop a sieve theory based on the classification methodology and the properties of the prime numbers with respect to the two equations that have been developed.

2 The Concept of *G*-Numbers

In order to clarify what is happening within the sieve, another concept is brought in. Consider the following numbers in the set {2, 3, 5}. Let us define this as a base set to generate other elements, where this is based on the simple rule that whatever a new element is added, it must be a multiple of the base set. Then we have,

- {4, 6, 10} from 2
- {6, 9, 15} from 3
- {10, 15, 25} from 5

This gives us the effective set {2, 3, 4, 5, 6, 9, 10, 15, 25} with the new elements on the basis of the base set. If the base set is the first wave of numbers, then this becomes the second wave of numbers. Now to generate the third wave, we do the same operation again. The new elements are now generated on the basis of the set {4, 6, 9, 10, 15, 25}

- {12, 18, 20, 30, 50} from 2
- {18, 27, 30, 15, 75} from 3
- {24, 36, 40, 60, 100} from 4
- {30, 45, 50, 75, 125} from 5
- {36, 54, 60, 90, 150} from 6
- {60, 90, 100, 150, 250} from 10
- {90, 135, 150, 225, 375} from 15
- {100, 125, 150, 225, 250, 375, 625} from 25

This gives us the effective union set:

{2, 3, 4, 5, 6, 8, 9, 10, 12, 15, 16, 18, 20, 24, 25, 27, 30, 36, 40, 45, 50, 54, 60, 75, 81, 90, 100, 125, 135, 150, 225, 250, 375, 625}

THE THEORY OF PRIME NUMBER CLASSIFICATION

The above are referred to as G-numbers in the context of the following definition. This is a similar operation to the cross product.

Definition 1

> Let there be a base set $\{b_1, b_2 \ldots b_n,\}$ that produces new elements that enlarge the base set, where these are defined by some operation $\Omega(n)$ repeated n times. The consequent set is called the set of G-numbers.

In our example, Ω is defined as a multiplicative operation, but with the prime number theory, we use $6x + 7$ and $6x + 11$. We observe the following about this set using the example described.

1. The base set is enlarged to infinity if $n = \infty$. This is unlike a cross product that produces a new but independent set.
2. The G-numbers have a random behavior characteristic for a relatively large n.
3. Generated subsets of b_i are underlying sets, and the union represents the universal set defined by the base set.
4. The gap between the G-numbers is increasing as we generate more elements of the set.
5. The base set decides what elements are going to be included in the generated set of numbers.
6. In the example, the base set acts as a sieve to include numbers divisible by 2, 3, and then 5. Numbers divisible by 7 for example are excluded.
7. The square of the elements is a determining factor in how fast the gap between the numbers is increasing.
8. The frequency curve is similar to that of the prime numbers, and it is evident that more generation will generate a closer match.

It is necessary to develop the concept of the G-numbers in order to be able to understand the emphasis of the sieve theory developed here. The reason is that prime numbers are generated numbers, and in their generation, they are not alone. The notion used to describe the numbers that accompany them in their generation is the word "pseudoprime," which has been used in the same context here.

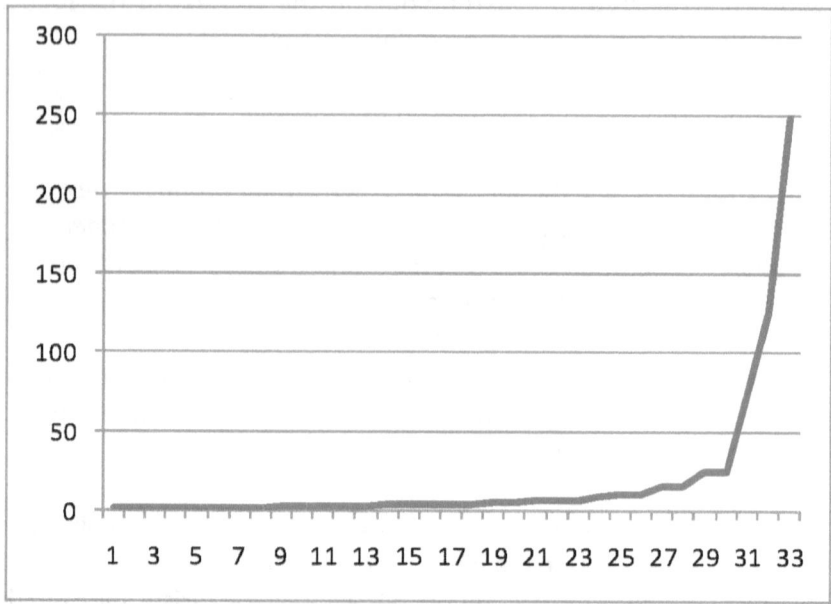

Figure 50. Frequency curve for G-numbers.

However, as the word "pseudo" indicates, it implies that the other numbers are not really the key; they are sort of accidental but necessary for consideration. In applying the concept of G-numbers, that type of emphasis in the relation between prime numbers and pseudoprimes is changed; primes and pseudoprimes are actually equals is the first assumption, and it is even possible that primes are the result of pseudoprimes as the second assumption.

3 Conceptual Structure of the Sieve

The concept of the sieve structure is therefore heavily based on the assumption of the definition of G-numbers. We first of all make the following assumptions:

Sieve Axioms

Let a grid called a G-set be defined where,

a. prime numbers behave according to the multivalued function $f(x) = (y_1, y_2)$;
b. all prime numbers exist either on $y = 6x + 7$ or $y = 6x + 11$, where $-\infty \leq Int\, x \leq \infty$; and
c. the pseudoprimes form the set of first G-numbers.

The prime numbers are alos regarded as G-numbers because they are generated. However, the pseudoprimes are regarded as the first G-numbers because they have a founding set. Now in the sieve the primes are taken to fill the space left by the pseudoprimes, hence the notion of the second set of G-numbers of the sieve being the primes.

It is typical to consider the prediction of a prime number to be dependent upon a single-valued function. This is how most of analytical analysis has developed, and it also influenced the development of number theory from that perspective. The attempt therefore has been to move away from the standard approach and try other methods of describing relationships. A key is the fact that we do not think of primes as occurring as a sequence in one set, rather they occur as a sequence in two sets.

3 CONCEPTUAL STRUCTURE OF THE SIEVE

Therefore, the grid is a conceptual construct in which the prime numbers exist, and they do not exist alone. This is the other major assumption, that prime numbers in this grid coexists with special pseudoprimes that are also related to the prime generation process rather than being a side event. The above assumptions have important implication that,

    a. all the prime numbers now have a specific position in a grid described by the sieve axioms,
    b. each specific position has a known value that can be found in the grid, and
    c. the grid consists of primes and a special set of pseudoprimes.

Hence, we predict a prime in terms of its position in the grid rather than describing it in terms of trying to define an nth prime. That is, we are not looking for the nth prime, rather we are looking for the value of a given prime at the nth position of the grid. Therefore, when we can establish a relationship between these two, we effectively develop the theory to predict where we can find a prime number on the grid and the value of that prime number. From that, we can then work backward to derive the sequential position of the prime number if we need to know that.

The immediate benefit of this approach is a primality test called the P1-test. For example, if we suspect that p is a prime number, then,

1. $x_1 = (p-7)/6$ or $x_2 = (p-11)/6$ must be a positive integer. If none is an integer, then immediately p is not prime.
2. If x_1 is an integer, then p is y_1. Similarly, if x_2 is an integer, then p is y_2. Only one of these can be true.
3. If one of them is true, then p is either a prime or pseudoprime.

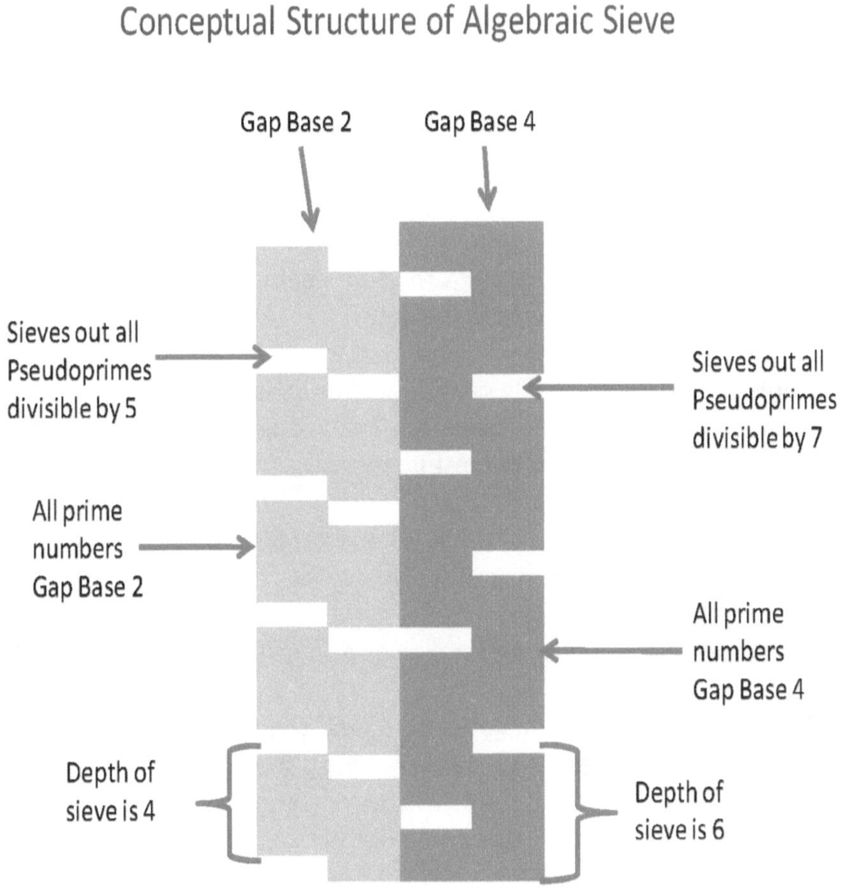

Conceptual Structure of Algebraic Sieve

Gap Base 2

Gap Base 4

Sieves out all Pseudoprimes divisible by 5

Sieves out all Pseudoprimes divisible by 7

All prime numbers Gap Base 2

All prime numbers Gap Base 4

Depth of sieve is 4

Depth of sieve is 6

Figure 51. The xy-sieve example.

The sieve as a construct is assumed to be a representation of a pseudoprime occurrence in gap base 2 or gap base 4, where such an occurrence is controlled by the sieve depth. Conceptually, using the fundamental gap theorem, then

a. the first part of the sieve is called the x-sieve, and
b. the second part of the sieve is called the y-sieve.

The numbers between the gaps in a sieve define a probabilistic occurrence of a prime number.

3 CONCEPTUAL STRUCTURE OF THE SIEVE

This implies the sieve is based on gap base 2 and gap base 4 structure. Consequently, this is referred to as an xy-sieve with the following properties:

1. Each prime number and pseudoprime defines an xy-sieve; hence, there are an infinite number of sieves. Therefore, each sieve is defined by $f(x) = (y_1, y_2)$.
2. Let there be a sieve defined by y_1, then the sieving process will be defined by a sieve step of y_1, the same being true for y_2.
3. Each sieve step defines a sieve depth.

An example of the grid for $x = 55$ to $x = 72$ has been given, where x denotes the position of either a prime or pseudoprime. We also note that x can denote the position of two primes or two pseudoprimes. That is why the concept of sequencing was revised because x can present two primes that are in sequence rather than one prime at a time. What is important is that these positions are uniquely related to the defined grid.

It is then observed that in the grid, taken as a snapshot:

a. All twin primes start in gap base 4, and the next prime will be in gap base 2. For example (419, 421) are twin primes, and 419 is in gap base 4, and 421 is in gap base 2. This gives some inspiration for handling the twin-prime conjecture challenge because it provides a consistent structure.
b. All pseudoprimes are divisible by either a prime or pseudoprime to give an integer.
c. There are no numbers that are divisible by 2 and 3 in the grid.

THE THEORY OF PRIME NUMBER CLASSIFICATION

x	6x+7	6x+11	Div by 5	Div by 5		Div by 7	Div b y 7
			Prime Numbers occupying fixed positions				
			Pseudoprime divisible by 7				
			Pseudoprime divisible by 5				
			Pseudoprimes divisible by either 11, 13, 17				
55	337	341	67.4	68.2		48.143	48.71429
56	343	347	68.6	69.4		49	49.57143
57	349	353	69.8	70.6		49.857	50.42857
58	355	359	71	71.8		50.714	51.28571
59	361	365	72.2	73		51.571	52.14286
60	367	371	73.4	74.2		52.429	53
61	373	377	74.6	75.4		53.286	53.85714
62	379	383	75.8	76.6		54.143	54.71429
63	385	389	77	77.8		55	55.57143
64	391	395	78.2	79		55.857	56.42857
65	397	401	79.4	80.2		56.714	57.28571
66	403	407	80.6	81.4		57.571	58.14286
67	409	413	81.8	82.6		58.429	59
68	415	419	83	83.8		59.286	59.85714
69	421	425	84.2	85		60.143	60.71429
70	427	431	85.4	86.2		61	61.57143
71	433	437	86.6	87.4		61.857	62.42857
72	439	443	87.8	88.6		62.714	63.28571

Figure 52. The grid showing the division by 5 and 7.

e. The occurrence of pseudoprimes is regular in the grid. An example is given using the primes 5 and 7, where these divide each column respectively.

When dividing by five, we call that a 5-sieve; by seven, a 7-sieve; by twenty five, a 25-sieve. So the sieve consists of a set of smaller sieves within a general framework.

A clear pattern emerges if we extract the occurrences of the pseudoprimes, where the shading indicates the prime numbers and the gaps the pseudoprimes. The interpretation code for the structure is given by the following table. Decimal and integer imply a division operation that results in a decimal or integer after division by $5, 7, 11, \ldots$ or any prime and pseudoprime.

Result	Result	Classification
Decimal	Decimal	Prime
Decimal	Integer	Pseudoprime
Integer	Decimal	Pseudoprime
Integer	Integer	Pseudoprime

Table 27. Classification Code for xy-sieve.

This pattern is then used to now develop the structure into an algebraic sieve that can give us the location of the prime number as well as its value. The mechanics of the sieve are as follows:

1. When dividing by a prime or by a pseudoprime, a decimal indicates a probability of a prime. For this reason, when we apply the interpretation code, we need to divide the same number both by 5 and 7 for example. This explains the use of the two columns for each of the divisors. The first column is said to define the x-sieve, and the second column defines the y-sieve.
2. Since the pseudoprimes are regular, we focus on them in order to identify the prime numbers in the sieve. They form the basis of formulating the algebraic sieve.
3. A [decimal, decimal] occurrence on both the x-sieve or y-sieve confirms a prime number. For example, at $x = 7$, we have 49 and 53. The $49 = [9.8, 7]$, and $53 = [10.6, 7.57143]$, which means according to the interpretation table, we have a pseudoprime and a prime number.

One may be tempted to immediately conclude that there is nothing new here really because the approach uses the same elimination technique of dividing to remove divisible numbers so that the ones that remain are prime. Indeed, this is a common ground, the major difference is as follows:

a. We only divide by numbers contained in the sieve, no other numbers are used. For example, we do not divide by two or three to determine a prime number. This is an intrinsic property of the sieve, that some numbers for division are already eliminated.
b. Secondly, we also divide by pseudoprimes, they are part of the process of identifying the prime number.

THE THEORY OF PRIME NUMBER CLASSIFICATION

c. The xy-sieve establishes a principle of regularity of pseudoprimes that defines the position of the next prime.
d. The sieve is based upon a special type of function called the multivalued function $f(x) = (y_1, y_2)$. Other sieves may not necessarily have a specific mathematical function that defines them. Consequently, it is referred to as an algebraic sieve.
e. At first, the foundational formula for prime generation were based on $p = 6a \pm 1$. This meant that we consider the first "a" as $a = 1$, giving us the primes 5 and 7. However, using sieve axioms, we introduced the multivalued function. This means a broader approach can be used as this allows the condition $-\infty \leq Int\, x \leq \infty$ to define prime generation in the context of the sieve.
f. It will also be demonstrated that the sieve explains why a formula for finding primes will work for a few values of n and then fail to predict the next value.
g. Observe that the xy-sieve brings out a special relationship between the primes {1, 5, 7, 11, 13, 17, 19, 23, 29}. There is no pseudoprime between these numbers if we take them according to their gap base.

Note that for prime generation, the first value of x is $x = 1$. However, for the sieve model, the first value of x is $x = -1$ for the natural primes.

X	6x+7	6x+11	Div-5	Div-5	Div-7	Div-7
-2	-5	-1	-1	-0.2	-0.71429	-0.14286
-1	1	5	0.2	1	0.14286	0.71429
0	7	11	1.4	2.2	1	1.57143
1	13	17	2.6	3.4	1.85714	2.42857
2	19	23	3.8	4.6	2.71429	3.28571
3	25	29	5	5.8	3.57143	4.14286
4	31	35	6.2	7	4.42857	5
5	37	41	7.4	8.2	5.28571	5.85714
6	43	47	8.6	9.4	6.14286	6.71429
7	49	53	9.8	10.6	7	7.57143
8	55	59	11	11.8	7.85714	8.42857
9	61	65	12.2	13	8.71429	9.28571

Figure 53. No pseudoprimes from 5 to 29.

4 Meaning and Interpretation

As explained in the foreword of this book, it was undertaken because of mathematical research into meaning and interpretation in mathematics. The most important aspect of a definition is the meaning that it imparts and, secondly, the resulting interpretation as a consequence of relationships with other existing mathematical objects. What is the essence of meaning? For expository mathematics, it is assumed that

1. conceptual consistency defines the essence of meaning in mathematical definition, and
2. the power to interpret meaning in other objects through mathematical relationships adds credibility to the consistency and applicability of the meaning conceived.

For example, before the advent of the negative numbers, the prime number definition was satisfactory in this manner:

A prime number is any number that is divisible by itself and one only

Some people still give this as a valid definition. With the advent of the natural number concept, the definition had to develop a little but further as shown.

A prime number is any natural number that is divisible by itself and one only.

In the above, we are mostly considering conceptual consistency of the idea of defining a prime number. However, when we relate this to other mathematical

THE THEORY OF PRIME NUMBER CLASSIFICATION

objects such as the negative numbers, we realize some inconsistencies. That is,

- a prime number p is divisible by minus one, and
- a prime number p is divisible by $-p$.

Hence, this makes us to try and improve the definition in order to manage the interpretation of relationship between other mathematical objects. That is,

> *a prime number is a natural number that has two distinct natural numbers as divisors: itself and one only.*

The qualification is that the prime number is a natural number, and its divisors are also natural. Relative to the negative numbers, we are saying we do not consider them as divisors of prime numbers. However, this still does not prevent us from dividing by a negative number because doing so is still a very valid mathematical process. The other shortcoming is the fact that an even number 2 is also regarded as a prime number, where this is not consistent with the observation that all prime numbers are odd. Expository mathematics deals with such problems that arise from meaning and interpretation of concepts, ideas, and definitions. There is also the perennial problem on whether to consider one a prime number or not.

The most significant outcome of applying the sieve theory leads to the following postulate.

Postulate 1

a. *1 is a prime number*
b. *2 and 3 are not prime numbers*

This is merely a postulate based on the conceptual structure of the sieve; it is not based on the conventional definition of a prime number. What the sieve indicates is that 2 and 3 do not fit into the axiomatic pattern described for the primes, and we may accept it as an anomaly that is a consequence of basing the system on gap base 2 and gap base 4, whose pattern is structured around 6. The other option is to consider what the sieve is suggesting through the postulate. It solves the argument on whether 1 is a prime number or not but throws a new argument that actually, even though 2 and 3 have no factors besides themselves and one, they are actually not prime. It creates a research question as to what then is a prime number. A possible definition could be as follows.

Definition 2

The Prime Number Axioms

A prime number is any number p such that,

1. *it is an element of $f(x) = (6x + 7, 6x + 11)$, where $-\infty \leq Int(x) \leq \infty$;*
2. *it is divisible by one and minus one; and*
3. *it is divisible by itself, and by its magnitude.*

As stated before, it must be remembered that prime numbers were discovered before the concept of the real number existed, so they have been defined only to be natural numbers. That definition has probably made sense until now. This is where sometimes we are bound by tradition and norms that accompany certain mathematical conclusions coupled with trying to create a universal approach to defining a concept such that it fits the broad spectrum of mathematics.

On the other hand, there must be a justifiable reason for demanding a change of perspective in a given definition. The assumption is that the sieve axioms give us a context for suggesting a conceptual change in the way prime numbers are defined. The values are the conceptual consistency of the system and model designed to describe the prime numbers and the holistic relationship that is assigned to them through the concept of a *G*-number.

The implication of this definition is as follows:

1. The prime number axioms affirm the postulate that 2 and 3 are not prime numbers. That is, in this context, they fulfill only one condition for a prime number, not all of them. This is the condition of being divisible by one and themselves only.
2. That prime numbers should also include the condition of being divisible by minus one. Actually, all natural numbers are divisible by minus one,

THE THEORY OF PRIME NUMBER CLASSIFICATION

including prime numbers. The omission of minus one reflects the fact that the definition was made when negative numbers were nonexistent in mathematics. Therefore, the prime number 19 is divisible by -19 to give -1. We overide all of this by making the assumption that prime numbers are positive, but this does not imply that primes cannot be divided by negative number. The assumption therefore is that 19 is divisible by $\pm 1, 19$ and -19.

3. Prime numbers can also be negative. For example, -19 is divisible by one, by -19, and by 19 only.

These changes are made and accepted on the basis of conceptual consistency and recognizing that at times it may be necessary to remove certain limitations or conditions in order to introduce a different perspective whilst at the same time retaining some aspects of the previous thinking.

We can introduce the notion of systems in order to substantiate the definition framework for the prime number. The system can be devised on the following concept axioms.

a. There exists a given set of elements in D_1 and D_2.
b. The only operation allowed in the sets is division.
c. The result of that division must be an element within the set D_1 only or within the set D_2 only.
d. The result must be an integer.

These are called concept axioms because we want to form or justify a definition out of them. Now, let $D_1 = \{P, a, 1\}$. Then we have the following outcomes describing the system of the set.

Divide	By	Result
P	A	P
P	P	a
a	A	1

Table 28. The first system.

191

If, for example, $P = 10$ and $a = 5$, the the result is 2, and this does not satisfy the axioms. Therefore, we see that $a = 1$, and p must have no factors; hence, we have $D_1 = \{P, 1\}$ in order to satisfy the above system.

Similarly, let $D_2 = \{P, -P, a, -a, 1\}$. Then we have the following outcomes describing the system of the set.

Divide	By	Result
P	a	P
P	P	a
P	-a	-P
P	-P	-a
a	a	1
a	-a	-1

Table 29. The second system.

Hence, to satisfy the set conditions, then $a = 1$, and P must have no factors. Therefore, $D_2 = \{P, -P, 1, -1\}$. From the same set, we can also have the system based on -P.

Divide	By	Result
-P	a	-P
-P	P	-a
-P	-a	P
-P	-P	a
a	a	1
a	-a	-1

Table 30. The third system.

The above then provides a framework for justifying the prime number definition and developing the context of interpretation of the meaning. The definition is a

THE THEORY OF PRIME NUMBER CLASSIFICATION

consequence of a system approach based upon a set of axioms called concept axioms. This is in contrast to the definition of the prime number that is based on attributes of divisibility by one and itself only.

Lastly, the other "mathematical evidence" that supports the use of negative primes is the construction of the algebraic sieve. The sieving operation uses two equations. The first equation uses positive primes, and the second equation uses negative primes to perform the sieving operation. In other words, negative primes are part of the description of numbers of the sieve.

5 Internal Dynamics of the Algebraic Sieve

The xy-sieve can also be studied in terms of its internal dynamics. There is also a clear structure of relationship within the sieve besides the sieving out of numbers such that only what remains are primes and pseudoprimes. It has the following internal dynamics:

a. There is a difference between the x-sieve structure and the y-sieve structure. This is a horizontal difference and a diagonal difference.
b. The horizontal difference is consistently 4.
c. The diagonal difference is consistently 2 from the right.
d. The diagonal difference is consistently 10 from the left.
e. The downward difference is consistently 6.
f. The sieve can also be used to define the delta space, that is, the prime and its associated gap. The gap is established from the shadow prime.

Internal Dynamics in the Structure of Algebraic Sieve

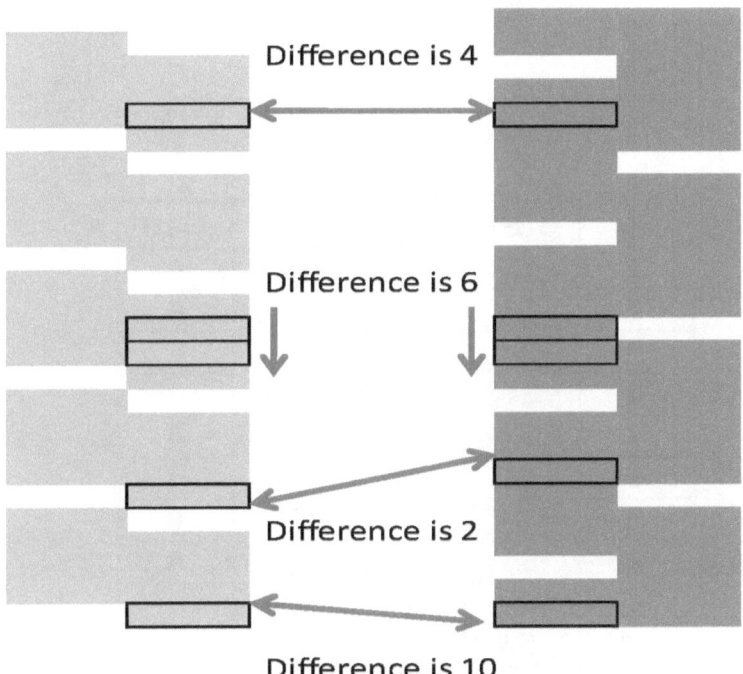

Figure 54. Internal structure of the xy-sieve.

The consequence of these dynamics is not really clear unless we try and apply the gap theory in terms of gap base. That is, we can assume as follows:

a. The horizontal difference of 4 is related to gap base 4, and this difference is also related to the prime number equations. That is, the difference in $f(x) = (y_1, y_2)$ for $y_1 - y_2 = 4$.

5 INTERNAL DYNAMICS OF THE ALGEBRAIC SIEVE

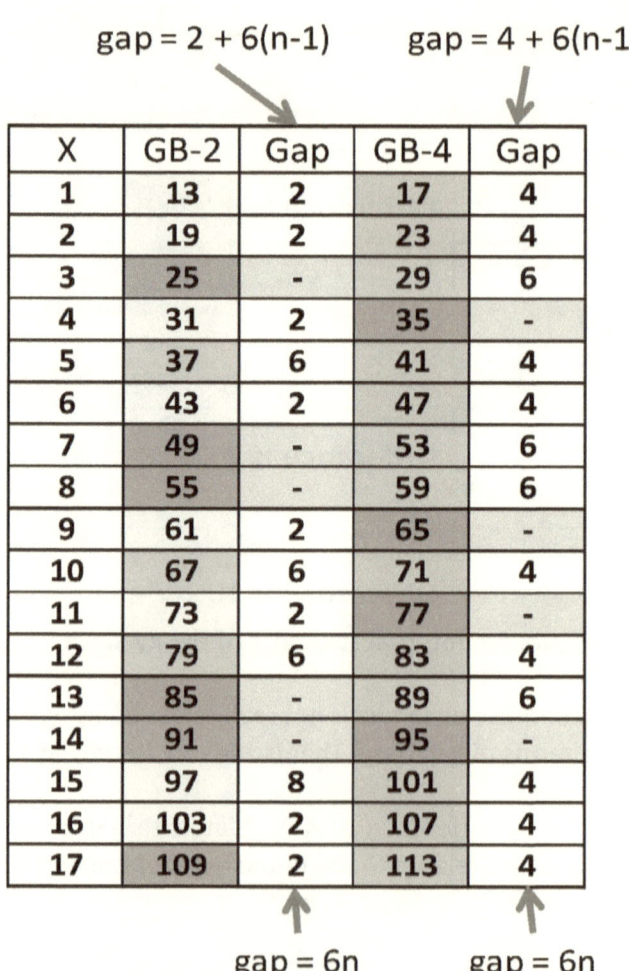

Figure 55. Illustrating the internal dynamics.

THE THEORY OF PRIME NUMBER CLASSIFICATION

   b. The diagonal difference from the right to the bottom left is related to the gap base 2. When both are prime numbers, this defines the twin primes.
   c. The downward difference of 6 is probably related to gap base 6 formation and also precisely related to the formation of the equations that generate the prime numbers. That is, it relates to the "$6a$" part of prime number generation.
   d. It is not clear what the difference of 10 relates to.

In terms of the natural gap of prime numbers, the internal dynamics of the xy-sieve tells us how the prime gaps will form between any two successive primes.

That is,

   a. the gap between any two successive primes will be 2. The first prime must be on the y-sieve, and the next prime must be on the x-sieve is the general rule.
   b. the gap between any two successive primes will be 4. The first prime number must be on the x-sieve and the next prime number on the y-sieve is the general rule.
   c. the gap between any two successive primes in the same gap base will be $6h$, where $(h-1)$ is the number of pseudoprimes between the two prime numbers.
   d. the gap between two successive prime numbers will be 10. The first prime number will be on the x-sieve, and the next prime number will be on the y-sieve is the general rule.

Through the sieve, the gap base theory also confirms the observation that 2 and 3 are not part of the prime family. If we include them in the sieve, then they produce gaps that are not consistent with gaps that are defined for the specific groups. This may be further evidence for arguing that they have the property of being divisible by one and themselves only, but this property is an anomaly with respect to the definition of the prime numbers.

5 INTERNAL DYNAMICS OF THE ALGEBRAIC SIEVE

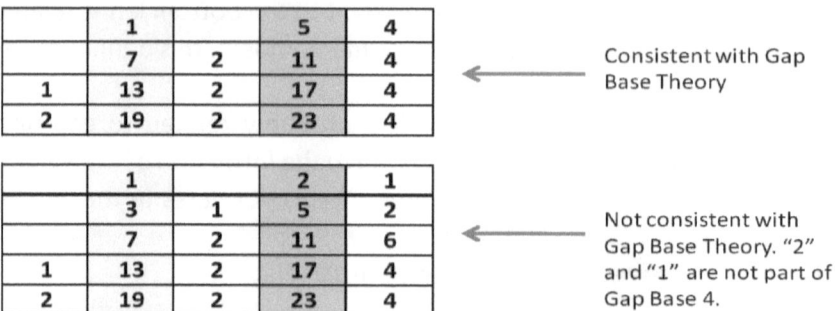

Figure 56. The different possibilities for the model.

Therefore, the algebraic sieve contains all the information that we need to know about prime numbers, and we put this statement as a postulate.

Postulate 2

The xy —sieve contains all the information in regard to gap formation and structure for prime numbers.

It is evidently clear then that this sieve is different from that of Eratosthenes.

5.1 PATTERN STRUCTURES

The internal dynamics therefore can be used to determine certain patterns of prime number occurrence. Normally, pattern structures in prime numbers are identified through the gaps between the prime numbers. Such patterns are based primarily on successive prime numbers, and famous amongst these is the twin prime, which defines a gap of two.

THE THEORY OF PRIME NUMBER CLASSIFICATION

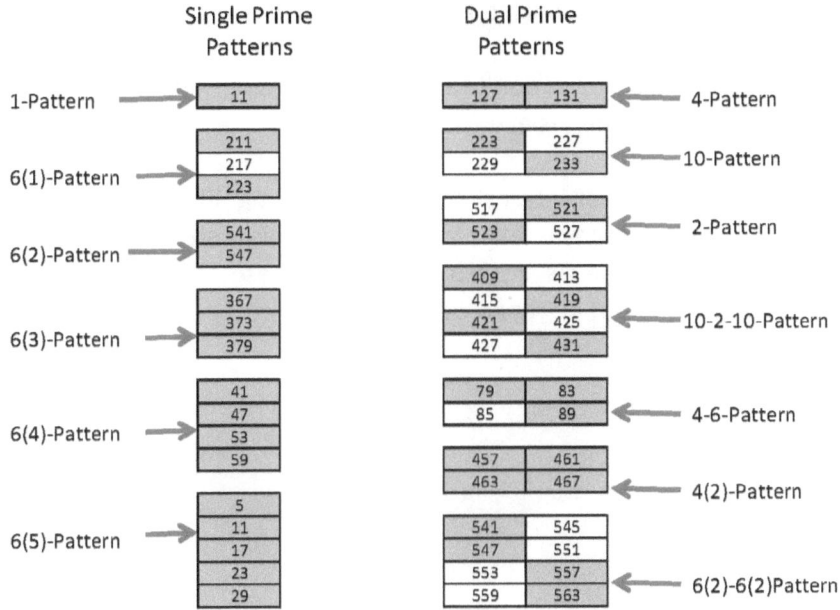

Figure 57. Sieve patterns of prime numbers.

The interesting outcome of the sieve application is that it also defines structures that reflect formation of prime numbers. Since they are specific patterns, they could also be a source of analytical and investigative interest.

In terms of the sieve, a prime pattern is defined in the following manner:

Definition 3

a. Patterns that occur on the x-sieve or the y-sieve are called single prime patterns.
b. Patterns that occur on both the x-sieve and the y-sieve are called dual prime patterns.
c. Any simplest pattern occurrence within a prime pattern defines pattern structure denoted by gap.

Therefore, gaps still play a prominent role in defining structure of patterns. The following will be noted:

1. All dual-pattern patterns start on the x-sieve with the exception of the 2-pattern and the 6-pattern.

2. The 2-pattern, also called twin-prime pattern, always starts on the y-sieve.
3. Some prime patterns don't seem to exist, for example, the 6-(4) pattern that starts on the x-sieve.

Therefore, it would be of interest to discover the significance of the prime patterns and also how they related to the pseudoprime structure. Because of the randomization of the sieve process, some of these patterns, for example the 6(5)-pattern, will not be repeated again.

6 The External Dynamics of the Algebraic Sieve

In terms of the external properties of the xy-sieve, we observe the following:

a. The primes are either on gap base 2 or gap base 4, that is, the x-sieve or the y-sieve.
b. The sieve identifies areas at which a "hit" will occur for a pseudoprime—it is called a pseudohit. This corresponds to nonoccurrence of a prime number.
c. For natural primes, prime generation varies from $-1 \leq x \leq \infty$, and as x varies, it defines a unique sieve through sieve depth. The sieve is considered for positive prime only but can do negatives as well.
d. Normally, prime numbers for the same x can be said to define the xy-sieve pair. Therefore (7, 11) form the same pair, and (5, 7) form a cross pair on the sieve.
e. The x-sieve has a regular pattern that is defined for each value of x, as x varies to infinity. For the x-sieve, the sieve depth is given by $s_1 = 6x + 6$.
f. The y-sieve has a regular pattern that is defined for each value of x, as x varies to infinity. For the y-sieve, the sieve depth is given by $s_2 = 6x + 10$.
g. The next pseudohit will always occur at (sieve depth + 1) from the previous hit, that is at $s_1 + 1$ of $s_2 + 1$.
h. Each sieve depth for a given x defines a possible occurrence of a prime number. The sieve depth has the same definition whether we have a cross pair or not.
i. The pseudoprimes, colored yellow in the diagram, indicate that these too have a sieve depth, which confirms the fact that are an integral part of the prime number-generation process.

6 THE EXTERNAL DYNAMICS OF THE ALGEBRAIC SIEVE

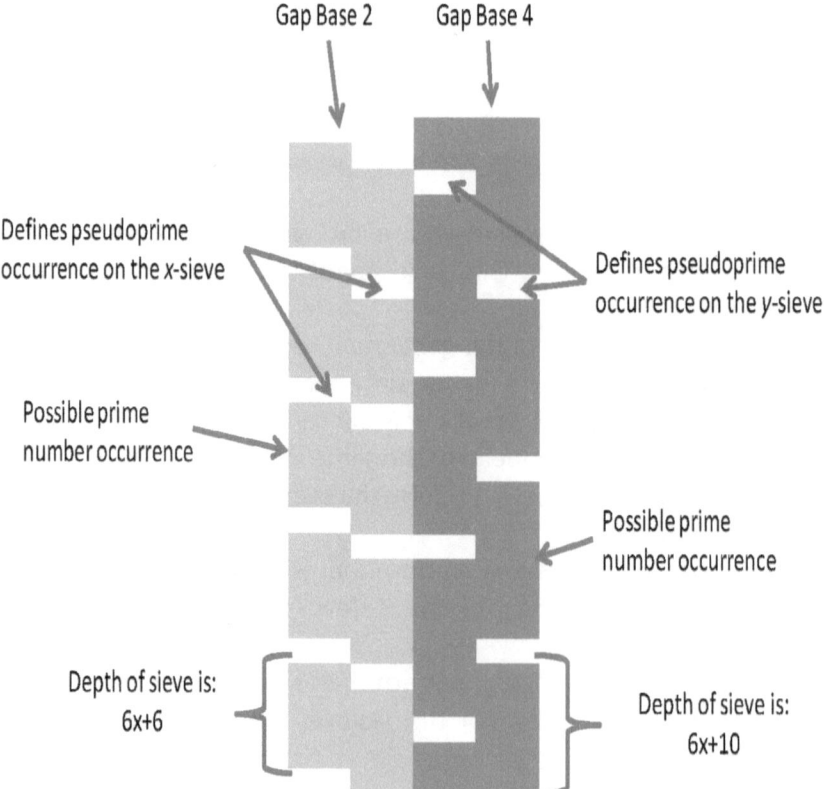

Figure 58. Structure of external dynamics of xy-sieve.

THE THEORY OF PRIME NUMBER CLASSIFICATION

The y-sieve depth for Gap Base 2

The x-sieve depth for Gap Base 4

x	6x+7	6x+11	6x+6	6x+10
-1	1	5	0	4
0	7	11	6	10
1	13	17	12	16
2	19	23	18	22
3	25	29	24	28
4	31	35	30	34
5	37	41	36	40
6	43	47	42	46
7	49	53	48	52
8	55	59	54	58
9	61	65	60	64
10	67	71	66	70
11	73	77	72	76
12	79	83	78	82
13	85	89	84	88

G-numbers

6 THE EXTERNAL DYNAMICS OF THE ALGEBRAIC SIEVE

Therefore, the algebraic description of the sieve is, as $-\infty \leq x \leq \infty$

$$\{x, 6x + 7, 6x + 11, 6x + 6, 6x + 10\}$$

For this reason, it is referred to as an algebraic sieve, where the depth is always one less the prime or pseudoprime. Hence, if $x = 3$, this defines the sieve

$$\{3, 25, 29, 24, 28\}.$$

This means we sieve out all numbers divisible by 25 and 29, and the depth of our sieve is going to be 24 and 28. The next sieve here will be $x = 4$, that is,

$$\{4, 31, 35, 30, 34\}.$$

Practically what happens is that one does a sieve on the x-sieve only and on the y-sieve only. That is, for example, the sieve $\{4, 31, 35, 30, 34\}$ will act on the x-sieve only, and then on the y-sieve only. Therefore, the operational assumptions and equations for a particular sieve are as follows, that is, one sieve acting in gap base 2 and gap base 4.

1. There exists on the sieve a point (x_1, y_1, y_2), y_1, y_2 is a prime or pseudoprime, and $-\infty \leq x_1 \leq \infty$, that is, x_1 can take any position on the sieve, where y_1 and y_2 defines a sieve.

2. Hence, for a given y_1 operating in gap base 2, then

    a. $x_1 = \frac{1}{6}(y_1 - 7)$ for the x-sieve

    b. $x_1 = \frac{1}{6}(-y_1 - 11)$ for the x-sieve on the y-sieve

3. Hence, for a given y_2 operating in gap base 4, then

    a. $x_1 = \frac{1}{6}(y_2 - 11)$ for the y-sieve

    b. $x_1 = \frac{1}{6}(-y_2 - 7)$ for the y-sieve on the x-sieve

4. For k an integer, then a "hit" in the sieve occurs whenever;

THE THEORY OF PRIME NUMBER CLASSIFICATION

a. $x = x_1 + ky_1$, at $(x, f(x))$, where $-\infty \leq k \leq \infty$, at initial point (a_1, a_2) on the x-sieve and and (a_3, a_4) on the y-sieve.

b. $x = x_1 + ky_2$, at $(x, f(x))$, where $-\infty \leq k \leq \infty$ at initial point (b_1, b_2) on the y-sieve and and (b_3, b_4) on the x-sieve.

Hence, for example, if we want to sieve all spots that are divisible by the prime number 13 in the x-sieve, where the x-sieve uses $f(x) = 6x + 7$, then;

a. $x = \frac{1}{6}(13 - 7) = 1$ for the x-sieve, and $f(x) = 13$ giving us the point $(1, 13)$ as our initial point.

b. Now $x = x_1 + ky_1$, hence $x = 1 + 13k, -\infty \leq k \leq \infty$.

c. Therefore, the equation for the x-sieve for 13 is $(1 + 13k, f(1 + 13k)), -\infty \leq k \leq \infty$.

Now for the y-sieve on the x-sieve, then

a. $x = \frac{1}{6}(-13 - 11) = -4$ for the y-sieve giving us the point $(-4, -13)$ as our initial point.

b. Now $x = x_1 + ky_1$, hence $x = -4 + 13k, -\infty \leq k \leq \infty$

c. Therefore, the equation for the y-sieve for 13 is

$(-4 + 13k, f(-4 + 13k)), -\infty \leq k \leq \infty$.

This will give all points from negative to positive infinity that are not occupied by a prime number and are pseudoprimes that are multiples of 13 on the x-sieve and x-sieve on the y-sieve. For $x = 1$, then $y_2 = 17$. We can find the sieve here too starting in gap base 4. For the y-sieve:

a. $x = \frac{1}{6}(17 - 11) = 1$ for the y-sieve giving us the point $(1, 17)$ as our initial point.

b. Now $x = x_1 + ky_2$, hence $x = 1 + 17k, -\infty \leq k \leq \infty$

c. Therefore, the equation for the y-sieve for 17 is

6 THE EXTERNAL DYNAMICS OF THE ALGEBRAIC SIEVE

$(1 + 17k, f(1 + 17k)), -\infty \leq k \leq \infty$.

Now for the y-sieve on the x-sieve, then,

a. $x = \frac{1}{6}(-17 - 7) = -4$ for the y-sieve on the x-sieve, and $f(x) = 17$ giving us the point $(-4, 17)$ as our initial point.

b. Now $x = x_1 + ky_1$, hence $x = -4 + 17k, -\infty \leq k \leq \infty$.

c. Therefore, the equation for the y-sieve on the x-sieve for 17 is

$(-4 + 17k, f(-4 + 17k))$, $-\infty \leq k \leq \infty$.

Hence, the general sieve equation is given as,

1. $(x_1 + y_1 k, f(x_1 + y_1 k)), -\infty \leq k \leq \infty$

2. $(x_1 + y_2 k, f(x_1 + y_2 k)), -\infty \leq k \leq \infty$

Note when $k = 0$, we have $(x_1, f(x_1))$, the starting point of the sieve. This means that we can make the sieve to be more useful by choosing the starting point and then let k vary over a narrow range rather than an infinite range. That is $k_1 \leq k \leq k_2$ for a given range $[a, b]$. Hence, for example, if we want to sieve between 6000 and 8000, we set k_1 and k_2 accordingly.

Hence, considering only integer values, we may expand the above equation as,

1. $(x + (6x + 7)k, 6\{x + (6x + 7)k\} + 7)$ for the x-sieve

2. $(\frac{1}{6}(-f(x) - 11) + f(x)k, -f(x) - 11 + 6(f(x)k) + 11)$ for the x-sieve acting on the y-sieve. Note $f(x) = 6x + 7$.

3. $(x + (6x + 11)k, 6\{x + (6x + 11)k\} + 11)$ for the y-sieve

4. $(\frac{1}{6}(-f(x) - 7) + f(x)k, -f(x) - 7 + 6(f(x)k) + 7)$ where this is for the y-sieve acting on the x-sieve. Note $f(x) = 6x + 11$.

THE THEORY OF PRIME NUMBER CLASSIFICATION

$x = 0$
This gives the 7-sieve and the 11-sieve

Legend:
- The x-sieve
- The x-sieve on the y-sieve
- The y-sieve
- The y-sieve on the x-sieve

Gap Base 2 and Gap Base 4 combined

7-sieve			11-sieve		
x	6x+7	6x+11	x	6x+7	6x+11
-5	-23	-19	-5	-23	-19
-4	-17	-13	-4	-17	-13
-3	-11	-7	-3	-11	-7
-2	-5	-1	-2	-5	-1
-1	1	5	-1	1	5
0	7	11	0	7	11
1	13	17	1	13	17
2	19	23	2	19	23
3	25	29	3	25	29
4	31	35	4	31	35
5	37	41	5	37	41
6	43	47	6	43	47
7	49	53	7	49	53
8	55	59	8	55	59
9	61	65	9	61	65
10	67	71	10	67	71
11	73	77	11	73	77
12	79	83	12	79	83
13	85	89	13	85	89
14	91	95	14	91	95
15	97	101	15	97	101
16	103	107	16	103	107
17	109	113	17	109	113
18	115	119	18	115	119
19	121	125	19	121	125
20	127	131	20	127	131
21	133	137	21	133	137
22	139	143	22	139	143
23	145	149	23	145	149

Gab Base 2 only

7-sieve	
x	6x+7
-5	-23
-4	-17
-3	-11
-2	-5
-1	1
0	7
1	13
2	19
3	25
4	31
5	37
6	43
7	49
8	55
9	61
10	67
11	73
12	79
13	85
14	91
15	97
16	103
17	109
18	115
19	121
20	127
21	133
22	139
23	145

Gap Base 4 only

11-sieve	
x	6x+11
-5	-19
-4	-13
-3	-7
-2	-1
-1	5
0	11
1	17
2	23
3	29
4	35
5	41
6	47
7	53
8	59
9	65
10	71
11	77
12	83
13	89
14	95
15	101
16	107
17	113
18	119
19	125
20	131
21	137
22	143
23	149

Figure 59. The sieving process.

6 THE EXTERNAL DYNAMICS OF THE ALGEBRAIC SIEVE

These are the sieving equations using x and k to identify the pseudoprimes. The smallest value of k is determined by the start of the range and the biggest range by the end of the range. Let the lower limit be defined by k_1 and the upper limit by k_2, both of which are integers. Hence, we get,

$$k_1 \leq Int \frac{a-7-x}{6x+7}$$

set by the value of a, and,

$$k_2 \geq Int \frac{b-7-x}{6x+7}$$

set by the value of b. This will sieve on the x-sieve for the given range; therefore, for the y-sieve, we have for the range a to b,

$$k_1 \leq Int \frac{a-7-x}{6x+11}$$

and

$$k_2 \geq Int \frac{b-7-x}{6x+11}$$

Therefore, for a given range $[a, b]$, we have the lower and upper bounds as,

$$k_1 \leq k \leq k_2$$

Now each x defines a sieve, but we do not need all the sieves because some of them will lie outside the range we have set as $[a, b]$. Since the limits for k_1 and k_2 can be determined, the next question is, how many sieves are needed to identify all the pseudoprimes in the given range? The number of possible sieves for the range is given by

THE THEORY OF PRIME NUMBER CLASSIFICATION

Sieving Equation:
$(x + (6x + 7)k, 6[x + (6x + 7)k])$

a =	40		k1	2
b =	100		k2	7
y(a)	247	Lower range		
y(b)	607	Upper range		
x	1	x remains constant to define the sieve		
k	1	k varies from 2 to 7		
sieve	13	This is the sieve defined by x = 1		
x-hit	y-hit			
14	91	This is the x and y hit as k varies between k1 and k2		

k	2	3	4	5	6	7	8
x-hit	27	40	53	66	79	92	105
y-hit	169	247	325	403	481	559	637

Figure 60. The 13-sieve.

Sieving Equation:
$(x + (6x + 7)k, 6[x + (6x + 7)k])$

a =	40		k1	1
b =	100		k2	4
y(a)	247	Lower range		
y(b)	607	Upper range		
x	2	x remains constant to define the sieve		
k	3	k varies from 1 to 4		
sieve	19	This is the sieve defined by x = 2		
x-hit	y-hit			
59	361	This is the x and y hit as k varies between k1 and k2		

k	1	2	3	4	5	6
x-hit	21	40	59	78	97	116
y-hit	133	247	361	475	589	703

Figure 61. The 19-sieve.

1. $6x_b + 7 = b$ for an x-sieve, hence $x_b = Int\left(\frac{1}{6}(b-7)\right)$
2. $6x_b + 11 = b$ for a y-sieve, hence $x_b = Int\left(\frac{1}{6}(D-11)\right)$

Let $c = x_b + 1$ the maximum number of sieves, then $0 \leq x \leq c$ for the range [a,b]. We need only use the first x_b as it is slightly larger, that is, the one from $6x + 7$. Therefore, the larger the range, the greater the number of sieves.

An example is given for [a,b] = [40, 100].

7 The Primality Test

A simple primality test was developed called the P1-test. This test only went as far as finding out whether we have a number that is part of the G-set or not. If it is part of the G-set, then it's either a prime of pseudoprime. What was left undone in this set was a way of determining whether we have a prime or pseudoprime once the P1-test has been done. This is constructed as follows, where this is called the P2-test.

In the sieving process, we noted that x defines the sieve, and k defines the sieving process while x is held constant. Hence for the P2-Test, we reverse the process, and we set x as a variable while we look for a corresponding k.

> Step One: Enter the number that has to be tested; say p
>
> Step Two: The test with setting this p to $6x + 7$ or $6x + 11$. If both are a decimal, the number is not a prime number. This is the end of the P1-Test. This is a very fast composite test. If one of them is an integer, then it is either a prime or pseudoprime, and we proceed to the P2-Test.
>
> Step Three: If $6x + 7$ gave an integer, then p is an element of gap base 2. However, if $6x + 11$ is an integer, then p is an element of gap base 4. Both cannot be an integer simultaneously in the two equations. This step determines which equations to use to determine the primality as shown in the table. Step 3 should derive t, which we hold constant while we vary x.

Comment	Gap Base 2	Gap Base 4
If x is a decimal, then number is not prime. This is a composite test.	$x = \dfrac{p-7}{6}$ Set $x = t_1$	$x = \dfrac{p-11}{6}$ Set $x = t_2$
Then the x-sieve and y-sieve equations will give p	$(x + (6x+7))k = t_1$ This derives k_1	$(x + (6x+11))k = t_2$ This derives k_1
The y-sieve on x, and the x-sieve on y will give p	$(\tfrac{1}{6}(-f(x)-11) + f(x))k = t_1$ This derives k_2 $f(x) = 6x+7$	$(\tfrac{1}{6}(-f(x)-7) + f(x))k = t_2$ This derives k_2 $f(x) = 6x+11$

Table 31. Equations for primality.

Therefore, for the x-sieve (gap base 2), we have,

$$k_1 = \frac{t_1 - x}{6x+7}$$

and the y-sieve on the x-sieve, we have,

$$k_2 = \frac{t_1 + x + 3}{6x+11}$$

Similarly, for the y-sieve (gap base 4), we have,

$$k_1 = \frac{t_2 - x}{6x+11}$$

and the x-sieve on the y-sieve, we have,

$$k_2 = \frac{t_2 + x + 3}{6x+7}$$

The interpretation code for the decision on whether it's a prime or pseudoprime is as given in the table.

Test Type	t_1	t_2	Decision Outcome
P1-test (exclusion)	Noninteger	Noninteger	Not prime
P1-test (inclusion)	Integer	Noninteger	Prime /Pseudoprime
	Noninteger	Integer	Prime /Pseudoprime
P2-test	k_1	k_2	
	Integer	Not Integer	Pseudoprime
	Not Integer	Integer	Pseudoprime
	Decimal	Decimal	Prime

Table 32. Decision criteria.

Hence, t_1 and t_2 gives us the value of x that generates the number we want to test. Now we know that when k is an integer (that is, k_1 or k_2 are integers), this defines a pseudoprime hit. So the first thing the test should do is to look for pseudoprimes, and secondly, a pseudoprime failure implies a prime number success. Failing to find an integer value of k then implies that we have a prime number, and the probability is one if both k_1 and k_2 are less than one, since no positive integer value of k can be defined. This is the basis of the primality test.

As x varies, then possible sieves occur that correspond to a given k. The value of k decreases as x increases. That is, x = -1 gives us the greatest values of k_1 and k_2. Also, if $x = -1$, then k_1 and k_2 will always be an integer for gap base 2 and 4 respectively. Therefore, we only consider the k_2 value for gap base 2 and k_1 for gap base 4. This leads to the following condition called the primality condition.

Definition 4—Primality Condition

Let x vary to from minus one to z, then if at z, $k_1 < 1$ and $k_2 < 1$ in a given gap base, then p is a prime number.

There will always be a sieve that satisfies this condition. Using the number 133, we demonstrate a test.

THE THEORY OF PRIME NUMBER CLASSIFICATION

Number = 133		t1 = 21		
X	Sieve	k1	k2	Sieve
-1	1	22	4.6	5
0	7	3	1.142857	11
1	13	1.538462	1.470588	17
2	19	1	1.130435	23
3	25	0.72	0.931034	29

Table 33. A pseudoprime test.

The number is a gap base 2 number because $t_1 = 21$; hence, we look at k values in gap base 2. The highest values of k_1 and k_2 occur at x = -1, but we ignore $k_1 = 22$ while $k_2 = 4.6$ tells us that we don't have a pseudohit. The sieves being used are 1 and 5 respectively. Then $x = 0$ gives us a pseudohit with sieve 7, so we have a pseudoprime in 133. Now note that if the progression continued at $x = 3$, we have both k_1 and k_2 being less than one. That is, we test for a pseudoprime first, followed by a failure condition that allows a prime to exist. Hence, $x = 3$ here is false because k_1 or k_2 was satisfied first for the pseudoprime condition.

Using the prime 41 as an example, we see that the P1-test will give us a value of $t_2 = 5$. This imples that the number is an element of the G-set and is in gap base 4. At this stage, the information is that the number is either a prime or pseudoprime. Therefore, we then pass on to do the P2-test to do primality operation. This gives the following result.

7 THE PRIMALITY TEST

Primality Test Example

Enter number		41	
The sieve is	x	1	
	13		13
Gap 2, x-sieve	1		13
Gap 4, x-sieve on y-sieve	-4		-13
The sieve is		17	
Gap 4, y-sieve	1		17
Gap 2, y-sieve on x-sieve	-4		-17

x	6x+7	6x+11
-1	1	5
0	7	11
1	13	17
2	19	23
3	25	29
4	31	35
5	37	41

	t1	t2
P1-Test Finding a1, a2	5.67	5.000

	t1	t2
P2-Test k1=	0.35897436 *by x sieve*	0.23529412 *by y-sieve*
k2=	0.56862745 *by y-sieve on x*	0.692 *by x-sieve on y*

If both a1 and a2 are not integer, then number is not prime
if k1 **OR** k2 is an integer, then number is a pseudoprime.
if k1<1 **AND** k2<1, then number is a prime.

Figure 62. Example with prime 41.

THE THEORY OF PRIME NUMBER CLASSIFICATION

x	k1	k2
-1	1.2	7
0	0.454545	1.142857
1	0.235294	0.692308

Table 34. The k values for testing primality.

Note that $k_2 = 7$ is an integer. We ignore this when $x = -1$ for any k_2 as it will always be an integer. But k_1 tells us that it is not a pseudoprime, so we proceed to the next x. The pseudoprime condition is a failure until we get k_1 and k_2 being less than one, giving us a prime at $x = 1$.

Besides testing for primality, the test can also be used to study the relationship between patterns of prime numbers and the pseudoprimes. An example is given in the table below.

Number	GB	A1	A2	K1	K2	X	Prime Type
103	2	16	15.33333333	0.736842105	0.913043478	2	Prime
1003	2	166	165.3333333	12.69230769	10	1	Pseudoprime
10003	2	1666	1665.333333	238	151.7272727	0	Pseudoprime
100003	2	16666	16665.33333	0.666733337	0.999900015	3332	Prime
1000003	2	166666	166665.3333	0.666673333	0.99999	33332	Prime
10000003	2	1666666	1666665.333	128205	98039.41176	1	Pseudoprime
100000003	2	16666666	16666665.33	25920	25760.08501	106	Pseudoprime
1000000003	2	166666666	166666665.3	8771929.684	7246377	2	Pseudoprime

Figure 63. Using the primality test to look for patterns.

We start with a number 103, and we keep on adding a zero in between the one and three. The interesting pattern that emerges is that all the numbers belong to gap base 2, but in terms of type they can either be a prime number or pseudoprime. None of the numbers is a composite! You can contrast this observation with the following result.

Number	GB	A1	A2	K1	K2	X	Prime Type
201		32.3333333	31.66666667				Not a Prime
2001		332.333333	331.6666667				Not a Prime
20001		3332.33333	3331.666667				Not a Prime
200001		33332.3333	33331.66667				Not a Prime
2000001		333332.333	333331.6667				Not a Prime
20000001		3333332.33	3333331.667				Not a Prime
200000001		33333332.3	33333331.67				Not a Prime
2000000001		333333332	333333331.7				Not a Prime

Figure 64. Not a prime or pseudoprime example through primality testing

7 THE PRIMALITY TEST

In this case, none of the numbers are a prime or pseudoprime.

7.1 Why Pseudoprimes Are G-Numbers

The pseudoprimes were defined as the first G-numbers, elevating their importance to more than the prime numbers. The pseudoprimes have the following characteristics:

1. They are generated by the same function that defines the prime numbers
2. They are all odd numbers
3. They contribute to the structure of the algebraic sieve, and hence are related to the primes in terms of the position they take on the sieve.
4. Some can be symmetric in nature, and can be truncated to give a pseudoprime, where this is a similar property of some prime numbers.

The prime like properties can be noticed from the numbers 3454321 and 123454321.

Pseudoprime	Gap Base	Action
3454321	2	Truncate on the left
454321	2	None
123454321	2	Symmetric

Figure 65. Prime like properties of pseudoprimes.

We also note that the prime numbers have an x^2 effect that is typical of the G-numbers, hence creating a link between the two types of numbers. T This provides very strong evidence that prime numbers are part of a G-set. The property of G-numbers is that,

- they are self-generating, that is, there exists a base set that generates the rest of the elements;
- they are random in distribution;
- they are predictable in terms of the underlying set, that is, an equation can be used to define the underlying set;

THE THEORY OF PRIME NUMBER CLASSIFICATION

- they have a gap behavior that is growing as the number of elements increase to infinity; and
- the change in gap is influenced by the square of the elements contained by the generating sets.

Most of these attributes have been shown to exist in relation to the prime numbers. Careful observation of the sieve demonstrates the following. The table can be continued to infinity but is only done to show a particular trend.

Originating G -Number	Square of G -Number	X	Comment
1	1	$x = -1$	The minimum number of factors you need to describe a pseudoprime. This effectively means there are no pseudoprimes.
5	25	$x = 3$	This is the first new pseudoprime. All the pseudoprime numbers that follow can be factored using 5.
7	49	$x = 7$	All pseudoprimes that follow will now be factored minimum using 5 and 7.
11	121	$x = 19$	All the pseudoprimes that follow will be factored using minimum 5, 7, 11.

Table 35. Pseudoprime structure.

We then have the minimum rule stated thus, and this rule is similar to the rule that the number of prime factors needed to test primality being determined from the square root of p. The difference is that this also includes pseudoprimes.

Minimum Rule

The highest number of factors needed to sieve is determined by $f(x)^2$.

For example, between 1 and 5, we use 1 as the only factor. This means effectively there are no pseudoprimes in between the prime numbers, which is why they define the fundamental set. This is inclusive of 29 if we consider the fact that it is on gap base 4. Therefore, we need 1, 5, and 7 to sieve up to 100. The square of 100 is ten, and these are less than ten. The implication is that more factors for sieving implies a greater gap between the G-set elements.

The Law of Prime Numbers

a. The base for G-set that defines the xy-sieve is given by the fundamental set $\{1, 5, 7, 11, 13, 17, 19, 23, 29\}$.
b. All pseudoprimes are defined by $f(x) \cdot f(x)$, where

   I. $y_{1,x}(y_1), -1 \leq x \leq \infty$, in gap base 2
   II. $y_{1,x}(y_2), -1 \leq x \leq \infty$, in gap base 4
   III. $y_{2,x}(y_1), -1 \leq x \leq \infty$, in gap base 2
   IV. $y_{2,x}(y_2), -1 \leq x \leq \infty$, in gap base 4

c. The prime gap increases in proportion to $n \{f(x) \cdot f(x)\}$, where n is the number of pseudoprimes in range $[a, b]$.

The fundamental set is defined on the basis of the fact that there is no pseudoprime between the prime numbers. Gap base 2 ends at 19, but gap base 4 ends at 29. The primes are generated as a complementary set to the pseudoprimes according to this rule.

The G-set Rule

If a position in the xy-sieve is not occupied by a pseudoprime, it is occupied by a prime number.

This rule is the basis of the primality test using the algebraic sieve. An example of the law of primes is illustrated as in the table 36.

$y_{1,-1} = 5$ Constant at $f(-1) = 5$	Gap Base 2 $f(-1)(6x+7), -1 \leq x \leq \infty$	Gap Base 4 $f(-1)(6x+11)$, $-1 \leq x \leq \infty$
5	5 × 1 = 5	5 × 5 = 25
5	5 × 7 = 35	5 × 11= 55
5	5 × 13 = 65	5 × 17 = 85
5	5 × 25= 125	5 × 23 = 115

THE THEORY OF PRIME NUMBER CLASSIFICATION

5	5 × 31= 155	5 × 29 = 145
5	Continues to infinity	Continues to infinity

Table 36. Demonstrating the law of the primes.

We also observe that,

a. $y_{1,x}(y_2), - y_{1,x}(y_1), = 10$
b. $y_{2,x}(y_2), - y_{2,x}(y_1), = 2$

This demonstrates that the internal sieve properties are defined by these two equations. Note that because x goes to infinity, there are an infinite number of pseudoprimes of gap 10 and gap 2. The law also explains why we have to sieve on both sides of the xy-sieve. That is, we have $y_{1,x}(y_1)$ and $y_{2,x}(y_1)$ and similarly, $y_{1,x}(y_2)$ and $y_{2,x}(y_2)$ in order to highlight the pseudoprimes.

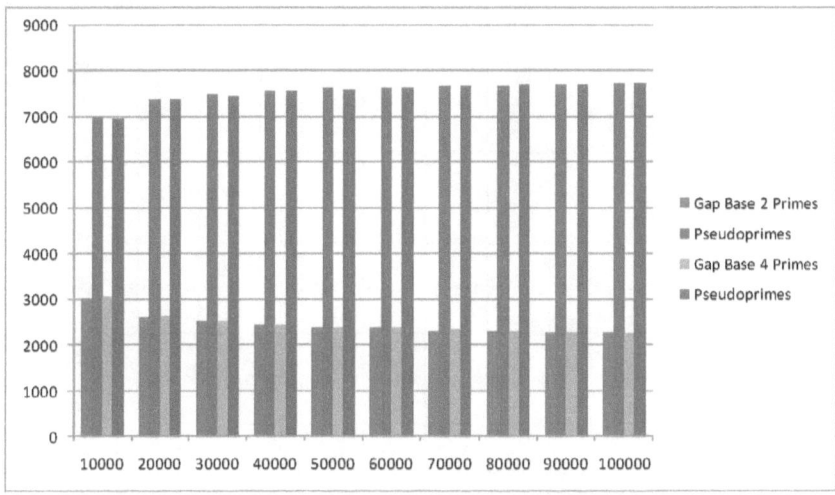

Figure 66. Number of primes and pseudoprimes almost the same per gap base.

7 THE PRIMALITY TEST

x	gb2	gb4											
-2	-5	-1											
-1	1	5	1										
0	7	11	1										
1	13	17	1										
2	19	23	1										
3	25	29	5										
4	31	35	5					5	5.8				
5	37	41	5					6.2	7				
6	43	47	5					7.4	8.2				
7	49	53	5	7				8.6	9.4				
8	55	59	5	7				9.8	10.6	7	7.571429		
9	61	65	5	7				11	11.8	7.857143	8.428571		
10	67	71	5	7				12.2	13	8.714286	9.285714		
11	73	77	5	7				13.4	14.2	9.571429	10.14286		
12	79	83	5	7				14.6	15.4	10.42857	11		
13	85	89	5	7				15.8	16.6	11.28571	11.85714		
14	91	95	5	7				17	17.8	12.14286	12.71429		
15	97	101	5	7				18.2	19	13	13.57143		
16	103	107	5	7				19.4	20.2	13.85714	14.42857		
17	109	113	5	7				20.6	21.4	14.71429	15.28571		
18	115	119	5	7				21.8	22.6	15.57143	16.14286		
19	121	125	5	7	11			23	23.8	16.42857	17		
20	127	131	5	7	11		24.2	25	17.28571	17.85714	11	11.36364	
21	133	137	5	7	11		25.4	26.2	18.14286	18.71429	11.54545	11.90909	
22	139	143	5	7	11		26.6	27.4	19	19.57143	12.09091	12.45455	
23	145	149	5	7	11		27.8	28.6	19.85714	20.42857	12.63636	13	
24	151	155	5	7	11		29	29.8	20.71429	21.28571	13.18182	13.54545	
25	157	161	5	7	11		30.2	31	21.57143	22.14286	13.72727	14.09091	
26	163	167	5	7	11		31.4	32.2	22.42857	23	14.27273	14.63636	
27	169	173	5	7	11	13	32.6	33.4	23.28571	23.85714	14.81818	15.18182	
28	175	179	5	7	11	13	33.8	34.6	24.14286	24.71429	15.36364	15.72727	
29	181	185	5	7	11	13	35	35.8	25	25.57143	15.90909	16.27273	
							36.2	37	25.85714	26.42857	16.45455	16.81818	

Gap expansion influenced by elements of G-set

Figure 67. Showing the square generation pattern.

THE THEORY OF PRIME NUMBER CLASSIFICATION

Let there be a range $[a, b, x]$, where such a range is constant at $(b - a)$ but varies its position in x. Since the sieve is a fixed structure at a given x, then at any point in the prime number distribution, the sum of the primes and pseudoprimes is a constant for the given range difference.

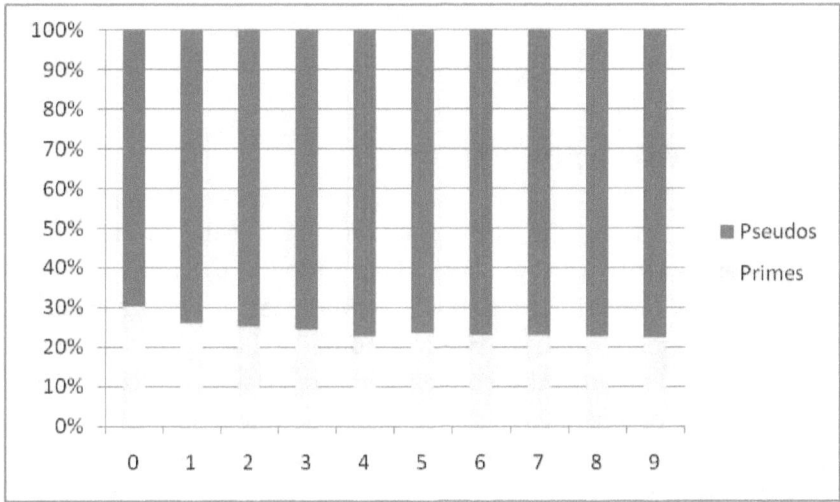

Figure 68. Relationship between primes and pseudoprimes (0-100000).

That is, if the number of primes is n_1 and the number of pseudoprimes is n_2, then the $n_1 + n_2 = c$, where c is a constant for the range difference $(b - a)$. Therefore, in a given range difference, the probability of getting a prime is $\frac{n_1}{c}$ and a pseudoprime $\frac{n_2}{c}$. Similarly, $n_1 < n_2$ for a reasonably large range. Now as x increases, the value of n_2 also increases. This then demonstrates that the generation of pseudoprimes affects the distribution of the prime numbers in an inverse relation. An example shows primes for $b - a = 10000$. Therefore, as $n \{f(x) \cdot f(x)\}$ increases, the number of primes is a given range decreases, and the gap between them increases.

8 The Algebraic Prime Space

Using the Law of Primes, the prime numbers are directly connected to the pseudoprimes in terms of their occurrence. They two numbers seemingly fight for the same space because they are connected algebraically, while the composite is independent from the prime number. This leads to the following definition that differs with the conventional approach. The context of the definition is in the current classification theory, not divisibility.

> **Definition 5**
>
> *Any number that is independent of the prime number distribution behavior is a composite.*

This implies three forms of numbers; primes, pseudoprimes, and composites. Let us assume that the number of composites for the range $[a, b]$ is n_3. Then it follows that, for a given gap base, we have the following.

$$f(b) - f(a) = n_1 + n_2 + n_3$$

That is the sum of primes, pseudoprimes and composites is defined by the given range. Now we know that $n_1 + n_2 = c$ where c is a constant. This value c is the same for both gap base 2 and gap base 4, therefore n_3 is the same for both gap bases. Similarly, for a given range $[a, b]$, then it means $f(b) - f(a)$ is also a constant.

This derives a space, an algebraic prime space. Actually the algebraic sieve represents the universal space, and therefore $[a, b]$ is a particular subspace.

THE THEORY OF PRIME NUMBER CLASSIFICATION

Definition 6

Let $G[a, b, c]$ be an algebraic prime space, and $c = n_1 + n_2$.

Hence the above then derives a probability space because our variables c and n_3 are fixed at any value of x for a given $[a, b]$. This derives the following probabilities of occurrence for a given a range $[a, b]$.

$$P(prime) = \frac{n_1}{f(b) - f(a)}$$

$$P(pseudoprime) = \frac{n_2}{f(b) - f(a)}$$

$$P(composite) = \frac{n_3}{f(b) - f(a)}$$

This approach gives us an alternative way of investigating the density of primes. This is typically given as an approximation given below.

$$\pi(x) = \frac{x}{Log\ x}$$

This is called the prime number theorem, and the notation $\pi(x)$ denotes the number of primes less than or equal to x. The range $[a, b]$ gives us an opportunity to study the prime density over a given range. Consequently, an abstract measure called the P-Index is defined that is related to the probability of occurrence of a prime, a pseudoprime, and a composite.

Definition 7

Let the probability of occurrence of prime, pseudoprime and composite in the range [a,b] be σ, then the P-index is given by 10000σ.

This will make the index to vary between zero and one hundred. For example if we define a range $x = 10$, and $x = 24$, that is $[a, b] = [10, 24]$, then $f(b) - f(a) = 84$. The table shows the total primes, pseudoprimes and composites, and the P-Index. Note that the range is kept constant, and the total number of primes plus pseudoprimes and composites remains constant. However, the P-Index is not constant because of the variation of the primes in the given range. Therefore it accurately reflects the density behavior of the prime numbers.

Range	Primes	Pseudoprimes	Total	Composites	P-Index
[10, 24]	16	14	30	138	65.1927
[19, 33]	16	14	30	138	65.1927
[60, 74]	15	15	30	138	65.4838
[560, 574]	8	22	30	138	51.2229
[1560, 1574]	12	18	30	138	62.8644
[11560, 11574]	9	21	30	138	55.0064

For a given $[a, b]$, one can search for a prime space using either c, or c and n_3. For example consider the space G[1,450,150]. This means the space has 150 primes and (450-150) = 300 pseudoprimes. When you perform a search to see is such a space exists, then [1348, 1797] as the exact space that suits the given condition. But if you seek a space G[1,350,150], you will get a nearby space with 130 primes and 220 pseudoprimes. What this means is that you cannot get the space you are actually looking for, but a nearby space is the closest you can come to considering the given conditions.

Whilst this may seem abstract, such a spatial concept can be applied to the market place, though the idea still has to be developed further. One can consider the following framework as a starting point, that is:

1. the prime numbers stand for your market share
2. the pseudoprimes stand for your competitor market share
3. the composites, which are independent, stand for the customers
4. The range [a,b] can be company resources or constraints

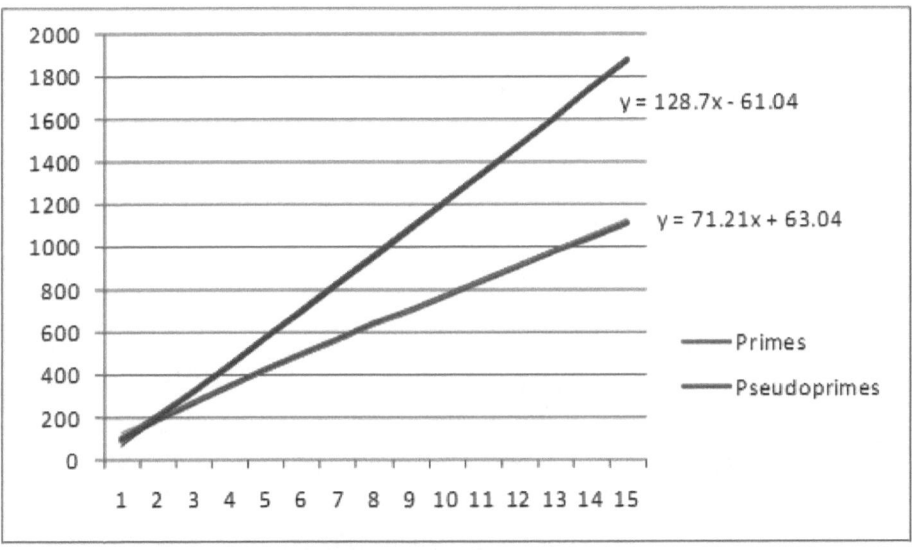

Figure 69. The primes and pseudoprimes in a cumulative space.

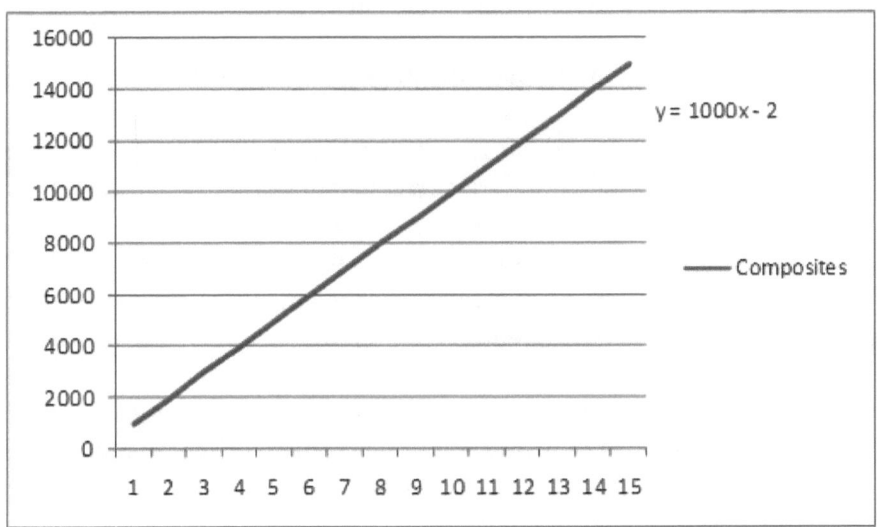

Figure 70. The composites in a cumulative space.

8 THE ALGEBRAIC PRIME SPACE

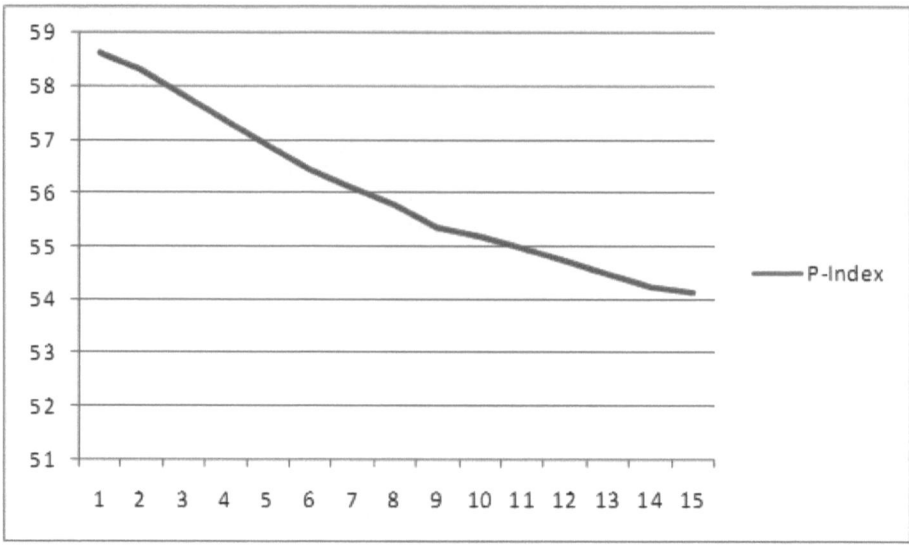

Figure 71. The P-index in a cumulative space.

The market, which is represented by c, can be taken to be a constant variable, including the customers within that market. The algebraic space therefore has all these characteristics. What is needed probably is to find a way of developing a theoretical model that can include the P-index.

The concept of the algebraic space is simple but revealing. There are many variations for studying the algebraic space, only two are given here as a demonstration. For example you can consider a space [0,100], and a next space [100, 200] and so on. This defines a constant space $[100n, 100n + 100]$. The other alternative is to consider a space [0,100], then [0,200] and so on. This defines a cumulative space $[0, 100n]$. In each case n is the number of subspaces The graphs that are shown are for the space $[0, 100n]$. It illustrates the following dramatic results, that:

1. The prime numbers is a given gap base increase in an almost linear relationship. It gives a very close approximate of the predictability of the density of primes as x increases.
2. The pseudoprimes in a given gap base increase in an almost linear relationship. Similarly, there is a close approximation on the predictability on the density of pseudoprimes.

THE THEORY OF PRIME NUMBER CLASSIFICATION

3. The composites increase in an almost linear relationship. Surprising the composite density is also almost linear in nature.
4. The P-index is a decreasing curve.

The primes and pseudoprimes are diverging as they increase, but interesting enough they have an intersection point. The scenario is completely different for the space $[100, 100n + 100]$, where n is the number of subspaces. Here the number of primes plus pseudoprimes is constant, and the number of composites is also constant. It is observed that:

1. the density of primes for the subspaces traces a curve.
2. the curve of the primes is a reflection of the curve of the pseudoprimes. This shows the close behavioral characteristic pattern of the primes and the pseudoprimes.
3. the P-index curve is much more erratic than the previous space definition.

One can also study the range [a,b] using the gap bases only.

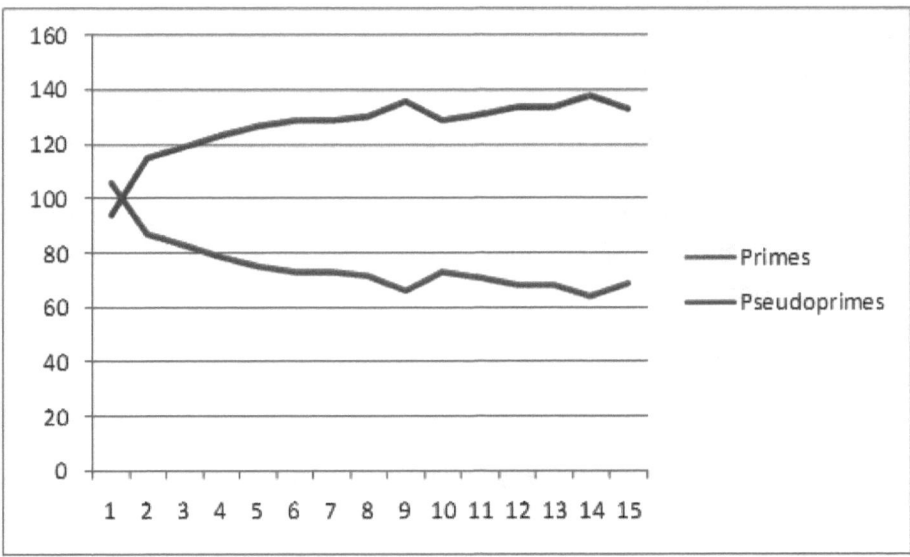

Figure 72. Prime and pseudoprime behaviour for a constant space

8 THE ALGEBRAIC PRIME SPACE

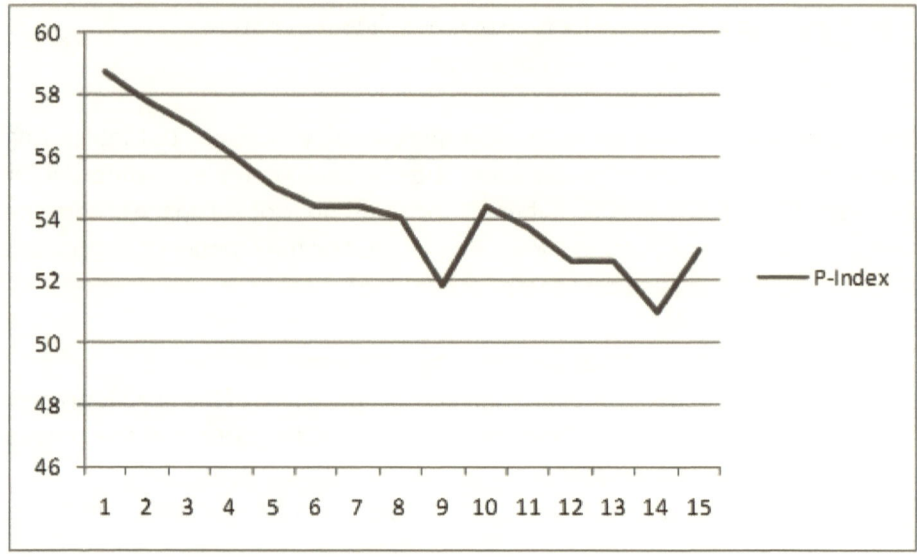

Figure 73. The P-index for a constant space

8.1 THE INTEGER PRODUCT LAW

Further interesting observations can be made in regard to product properties involving primes, pseudoprimes and composites. From the fact that the gap base structure was observed as a 2-4-6 sequence, by implication we can assume a sequence 0-2-4-6. That is, the set of composites is defined as gap base zero. Whereas 2-4-6 is repeatable as 8-10-12 and so on, the zero is not repeatable as a base. The implication is that such a set is unique in some way. This observation is summarised by the Integer Product Law.

Integer Product Law

If $p_1 \neq 1$ and $p_2 \neq 1$, then the product of any two numbers p_1 p_2 is either a composite in gap base zero, or a pseudoprime in gap base 2 or 4.

Normally, a composite has no gap base, and it is defined as any integer that has factors other than one and itself, or simply as any number that is not prime. But by considering them as a gap base, we can then state particular properties of such a gap base.

THE THEORY OF PRIME NUMBER CLASSIFICATION

The above rule can be elaborated further in the following manner.

1. [composite] x [composite] = [composite]
2. [composite] x [prime] = [composite]
3. [composite] x [pseudoprime] = [composite]

In terms of a product involving a composite, this is a closed set. As long as a composite is involved then the answer will be in the same set, the composite set. Composites define a gap base zero but are not considered to have random behaviour. Indeed, we can predict the set of all even numbers or odd numbers.

We further apply the law as follows to the *G*-set which consists of primes and pseudoprimes.

1. [prime] x [prime] = [prime], provided $p_1=1$ and p_2 is prime
2. [prime] x [prime] = [pseudoprime]
3. [prime] x [pseudoprime] = [pseudoprime]
4. [pseudoprime] x [pseudoprime] = [pseudoprime]

Hence if you multiply any two numbers, and the result is a pseudoprime, then there is a possibility that one of them could be a prime number. The law also emphasises the argument that the composite number should be treated separately from the pseudoprime. Note that 2 and 3 are composites, and as such, no product involving these two numbers will belong to the *G*-set. The following can also be derived in regard to the *G*-set.

1. If p_1 and p_2 are elements of gap base 2, then $p_1 p_2$ is an element of gap base 2
2. If p_1 and p_2 are elements of gap base 4, then $p_1 p_2$ is an element of gap base 4
3. If p_1 is an element of gap base 2, and p_2 is an element of gap base 4, then $p_1 p_2$ is an element of gap base 4

This means the rule is also binding on gap base 6, since this gap base is either a subset of gap base 2 or 4.

8.2 GENERATING GAP BASE ZERO

This does not imply that there is no gap between composites, rather it denotes the fact that this is a different class of numbers with respect to the primes and pseudoprimes. The set of all even numbers can be described by $2x$, while the set of all odd numbers can be described by $2n + 1$. However, some of the odd numbers are also prime or pseudoprime, hence we need to find a way of describing a relation for odd composites only. We begin by defining the gap base zero to consist of two families, family 1 being the odd numbers, and family 2 being the even numbers. In a way, this is similar to the G-set consisting of two sets, gap base 2 and 4.

Now we observe that all family 2 composites are generated by 2x to give an even number. We can identify odd composites by using $6x \pm 1$. When we divide the odd number by this equation, if we get a decimal for both conditions, then the number is composite. This is indicated in the diagram. This shows an interesting pattern, that family 1 composites are also regular in occurrence. Now if we plot the occurrences against x, we can establish a relationship with respect to n. The values of x being $1, 4, 7, 10, 13, \ldots$ are defined by the equation $3x - 2$. Hence, deriving the odd number composite generator is given by the equation;

$$2n + 1 = 2(3x - 2) + 1 = 6x - 3$$

Therefore we can now describe the family 1 in terms of x to give a table in the same format like gap base 2 and 4. We also note that there is a growing gap between the family 1 and 2 composites. This gap grows in step of four. This is contrast to the behaviour of gap base 2 and 4, which always have a constant difference.

We also note that there is a fixed pattern when we combine family 1 and 2 of the composites. This gives us a (1,1) and (2,2) pattern for gap difference to define two composite families. All of these observations lead to the following conjecture.

Axiomatic Assumptions on Composites

1. Composite are exactly predictable algebraically

THE THEORY OF PRIME NUMBER CLASSIFICATION

2. The gap between composite is 2 for even numbers, and six for odd composite numbers.
3. Between any two odd composites, there are either:
   a. two prime numbers
   b. or two pseudoprimes
   c. or a prime and a pseudoprime
4. There are always three even composites between any two odd composites

				is a composite		
	Even		Odd	Composite testing		
x	2x	n	2n+1	6x+1	6x-1	
1	2	1	3	0.333333	0.666667	1
2	4	2	5	0.666667	1	
3	6	3	7	1	1.333333	
4	8	4	9	1.333333	1.666667	2
5	10	5	11	1.666667	2	
6	12	6	13	2	2.333333	
7	14	7	15	2.333333	2.666667	3
8	16	8	17	2.666667	3	
9	18	9	19	3	3.333333	
10	20	10	21	3.333333	3.666667	4
11	22	11	23	3.666667	4	
12	24	12	25	4	4.333333	
13	26	13	27	4.333333	4.666667	5

Figure 1. The class 2 composite distribution in gap base zero.

8 THE ALGEBRAIC PRIME SPACE

A

x	Class 1 Even 2x	Class 2 Odd 6x-3	OE Gap 4x-3	Gap
1	2	3	1	
2	4	9	5	4
3	6	15	9	4
4	8	21	13	4
5	10	27	17	4
6	12	33	21	4
7	14	39	25	4
8	16	45	29	4
9	18	51	33	4
10	20	57	37	4
11	22	63	41	4
12	24	69	45	4

B

Gap Base Zero Composite Sequence	Gap Pattern
2	
3	1
4	1
6	2
8	2
9	1
10	1
12	2
14	2
15	1
16	1
18	2
20	2

Figure 2 The gap behaviour and pattern in gap base zero

Let group one of composites be those with a gap of one, and group two those with a gap of two. Therefore groups are defined in the same manner as in the case of prime numbers, that is, (3,1), (4,1), (9,1) (10,1) are composites in the same group. Therefore the composite groups are not confined to the respective families. Now, which is the first composite if we ignore the negative numbers. By looking at the gap pattern, it appears that zero must be the first composite, then we have the pattern as follows. Hence (0,2), (2,2), (6,2), (8,2) and so on are in the same composite group.

Composite Sequence	Gap Pattern
0	2
2	2
3	1
4	1

Figure 3. The gap pattern for numbers 0,2,3

THE THEORY OF PRIME NUMBER CLASSIFICATION

If we combine the arguments raised in gap base 2 and 4 about the numbers 2 and 3, we see that these numbers fit perfectly in the descriptive pattern of gap base zero. In other words, this is where they belong. All of these arguments are made in respect of pattern methodology and classification constructs.

For generation of the composites in terms of a classification system, consider the odd composites as a base class. We note that in the gap occurrence between composites, the occurrence of gap 1 is defined by $6x - 3$, which implies that we can use this as a base class. Now the base class plus one gives us the next composite with gap 1, and the base class plus three and five respectively gives us the composites where the gap difference is two.

	Family-1		Family-2	
x	6x-3	Plus 1	Plus 3	Plus 5
	Class 1A	Class 1B	Class 2A	Class 2B
0	-3	-2	0	2
1	3	4	6	8
2	9	10	12	14
3	15	16	18	20
4	21	22	24	26
5	27	28	30	32
6	33	34	36	38
7	39	40	42	44
8	45	46	48	50
9	51	52	54	56
10	57	58	60	62

Figure 4. The different composites classes

All the even composites define a family, called Family 2, and the odd composites define Family 1. Hence Family 2 has three classes, while Family 1 has only one class. However, class 1A and 1B defines group one composites, and class 2A and 2B define group two composites. When we combine all the generating formulae in terms of x, then this gives us the universal number distribution theorem as illustrated graphically. That is, we assume three families that described the distribution of numbers, where family-3 consists of numbers that are from gap base 2 and 4.

8 THE ALGEBRAIC PRIME SPACE

Theorem

Let x be an integer from negative to positive infinity, then all numbers are universally distributed as composite, pseudoprime and prime by $6x + y$, where $y = \{-3, -2, -1, 0, 1, 2\}$.

The implication of the theorem is that all numbers are algebraic, they can be expressed as an algebraic entity in relation to each other. Note that the composites behave like the prime numbers when the negative values are involved. That is, the positives of Class 1B become the negatives of Class 2B, but class 1A and class 2A positives are merely negated.

	Family-1	Family-2			Family-3	
x	6x-3	6x-2	6x	6x+2	6x-1	6x+1
	Class 1A	Class 1B	Class 2A	Class 2B	GB-4	GB-2
-8	-51	-50	-48	-46	-49	-47
-7	-45	-44	-42	-40	-43	-41
-6	-39	-38	-36	-34	-37	-35
-5	-33	-32	-30	-28	-31	-29
-4	-27	-26	-24	-22	-25	-23
-3	-21	-20	-18	-16	-19	-17
-2	-15	-14	-12	-10	-13	-11
-1	-9	-8	-6	-4	-7	-5
0	-3	-2	0	2	-1	1
1	3	4	6	8	5	7
2	9	10	12	14	11	13
3	15	16	18	20	17	19
4	21	22	24	26	23	25
5	27	28	30	32	29	31
6	33	34	36	38	35	37
7	39	40	42	44	41	43
8	45	46	48	50	47	49
9	51	52	54	56	53	55
10	57	58	60	62	59	61

Figure 5. The general distribution of numbers with "6" factor

THE THEORY OF PRIME NUMBER CLASSIFICATION

The factor of the difference of six is universal. The difference between all numbers along the column is six. We also note the following product particulars for positive numbers only..

Class	Class	Resulting Product Class
Class 1A	Class 1A	Class 1A
Class 1B	Class 1B	Class 1B
Class 2A	Class 2A	Class 2A
Class 2B	Class 2B	Class 1B
Class 1A	Class 1B	Class 2A
Class 1A	Class 2A	Class 2A
Class 1A	Class 2B	Class 2A
Class 1B	Class 2A	Class 2A
Class 2A	Class 2B	Class 2A
Class 1B	Class 2B	Class 2B
Class 1A	GB-2	Class 1A
Class 2A	GB-2	Class 2A
Class 1B	GB-2	Class 1B
Class 2A	GB-2	Class 1B
Class 1A	GB-4	Class 1A
Class 2A	GB-4	Class 2A
Class 2B	GB-4	Class 1B
Class 1B	GB-4	Class 2B
GB-2	GB-2	GB-2
GB-4	GB-4	GB-4
GB-2	GB-4	GB-2

Table 1: Product distribution

With the exception of Class 2B, all other products within a set form a closed operation. The distribution of the product class is probably a result of the algebraic structure of numbers.

The last observation that we may make is to note whether composites are random or not. From the behaviour pattern of prime numbers and pseudoprimes, it has been noted that structure in which these numbers exist is consistent and predictable, but the consistent structure does not imply predictable values within the structure. Hence the conclusion that prime numbers are random, the same being true for pseudoprimes. A similar argument holds for the composites. They exist in a consistent structure because there are precise equations to derive such composites, and the gap patterns are also consistent. However, all that this consistency does is to define a position in the given structure, but not the nth value of that position. Secondly, we work with the subsets in order to find values, but the universal set of composites has no equation to describe the value of a composite. This leads to a conjecture.

Conjecture

It is not possible to establish the value of the nth composite by a direct equation..

That is, regularity is positional and is a consequence of structure, but the actual value of the composite is independent of the position. The implication is that composites are also random, though they exist in a consistent structure.

9 Prime Number Curve

We can further apply the law of the prime by observing that the product between two value functions of x to give us the prime number curve. Knowing that we only consider integer values, then,

$$Int\{(6x+7)(6x+11)\} = Int\{36x^2 + 108x + 77\}$$

Now $6x+7$ and $6x+11$ are numbers in gap base 2 and gap base 4 respectively, and the horizontal difference between the two is always 4. That is,

$$(6x+11) - (6x+7) = 4$$

Therefore, prime numbers that have a difference of 4 will satisfy the quadratic equation to give us a prime number curve.

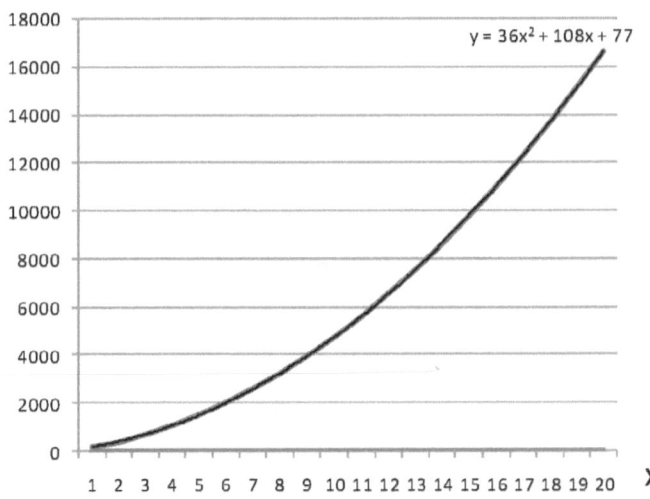

Figure 74. Product of y(1)y(2) for a given x.

But generally, we can consider the quadratic equation to give us a product between elements of the function $f(x) = (y_1, y_2)$ that is $(y_1)(y_2)$. That is, the consequence of the law of the primes is that all prime numbers are contained in the above equation through a product definition.

Type— y_1	Type— y_2	Gap
Prime	Prime	4
Prime	Pseudoprime	4
Pseudoprime	Prime	4
Pseudoprime	Pseudoprime	4

Table 37. The possibilities of the quadratic.

This can also be summarized as a theorem.

Theorem

Let there be consecutive primes p_1 and p_2 and consecutive pseudoprimes q_1 and q_2 in gap base 2 and 4 respectively, then,

a. $p_2 - p_1 = 4$

b. $q_2 - q_1 = 4$

c. $p_2 - q_1 = 4$

d. $q_2 - p_1 = 4$

and their product lies on $36x^2 + 108x + 77$.

The theorem provides a different way of expressing the relationship between the prime numbers and the pseudoprimes.

10 Conclusion 1—General Prime Generation

Note that the negative primes have switched in terms of their description using the gap base theory. The same positive prime in gap base 2 switches to gap base 4 when it is negative, an interesting observation. That is,

a. $f(x) = 6x + 7$ for $x \geq -1$ gives us gap base 2, positive;
b. $f(x) = 6x + 7$ for $x \leq -2$ gives us gap base 2, negative. However, this is the same as $f(x) = -6x + 11$ for $x \geq 2$ if we want to keep x positive.

In a similar manner,

a. $f(x) = 6x + 11$ for $x \geq -1$ gives us gap base 4, positive;
b. $f(x) = 6x + 11$ for $x \leq -2$ gives us gap base 4, negative. However, this is the same as $f(x) = -6x + 7$ for $x \geq 2$ if we want to keep x positive.

The above is an interesting property of the function $f(x)$ because it behaves like a directed function—values are being generated in a particular direction by the inequality. That is, for the elements of the same set, we can describe them in two ways as being true for $x \geq -1$ and for $x \geq 2$.

Now the function $f(x) = (6x + 7, 6x + 11)$ has been determined on the basis of the acceptable values of a_0 to establish the initial prime p_0 at $x = 0$. However, this is not the only function that generates primes since we know that we already have the function $6a \pm 1$ which was used for gap classification.

10 CONCLUSION 1 – GENERAL PRIME GENERATION

	Negative primes
	Positive primes

X	GB-2	GB-4
-11	-59	-55
-10	-53	-49
-9	-47	-43
-8	-41	-37
-7	-35	-31
-6	-29	-25
-5	-23	-19
-4	-17	-13
-3	-11	-7
-2	-5	-1
-1	1	5
0	7	11
1	13	17
2	19	23
3	25	29
4	31	35
5	37	41
6	43	47
7	49	53
8	55	59

Figure 75. The switch between positive and negative prime numbers.

THE THEORY OF PRIME NUMBER CLASSIFICATION

X	6X-37	6X-41	X	6X+7	6X+11	X	6X+1	6X+5
-7	-79	-83	-7	-35	-31	-7	-41	-37
-6	-73	-77	-6	-29	-25	-6	-35	-31
-5	-67	-71	-5	-23	-19	-5	-29	-25
-4	-61	-65	-4	-17	-13	-4	-23	-19
-3	-55	-59	-3	-11	-7	-3	-17	-13
-2	-49	-53	-2	-5	-1	-2	-11	-7
-1	-43	-47	-1	1	5	-1	-5	-1
0	-37	-41	0	7	11	0	1	5
1	-31	-35	1	13	17	1	7	11
2	-25	-29	2	19	23	2	13	17
3	-19	-23	3	25	29	3	19	23
4	-13	-17	4	31	35	4	25	29
5	-7	-11	5	37	41	5	31	35
6	-1	-5	6	43	47	6	37	41
7	5	1	7	49	53	7	43	47
8	11	7	8	55	59	8	49	53
9	17	13	9	61	65	9	55	59
10	23	19	10	67	71	10	61	65
11	29	25	11	73	77	11	67	71
12	35	31	12	79	83	12	73	77

Figure 76. The general prime number-generation functions.

10 CONCLUSION 1 – GENERAL PRIME GENERATION

Definition 4

Let G_p be a prime number-generating function, then

$G_p = (6z + k, 6z + k + 4)$ where,

a. $-\infty \leq Int\, z \leq \infty$

b. $k = f(x)$, and $\infty \leq Int\, x \leq \infty$, but k is constant for a given z.

Therefore, there are an infinite number of functions that generate the very same set. For our case, $k = 7$; hence, we have $6z + 7$ and $6z + 11$ as the generating functions. Further illustration is provided by figure 68, where k takes the values 1 and 37.

We have already stated that primes can be generated in the negative sense; however, we can also suggest that all numbers can be generated, provided we assume that x takes all values as the expression $-\infty \leq x \leq \infty$ correctly indicates. We have been only considering integers. However, we can also consider decimal values for x. For example

x	0	1.5	3	4.5	6	7.5	9	10.5	12	13.5	15	16.5	18	19.5	21	22.5	24
6x+7	7	16	25	34	43	52	61	70	79	88	97	106	115	124	133	142	151

In fact, taking all values of x will generate the set of all numbers. However, in terms of the concept of pseudoprimes, this would make it difficult to deal with prime numbers. We will be sort of back to square one.

11 Conclusion 2—The Sieve Format

A different approach could have been used for the sieve format depending on the choice of a_0 in the generating equation $p_n = p_0 + 6n$. Consider the table shown of possible values of derived using $6a_0 \pm 1$. As indicated before, the values of a_0 are paired, and the first value always starts from gap base 2, and a_0 must be an integer.

f(x)	GB-2 a_0	GB-4 a_0
1	0	0.333333
5	0.666667	1
7	1	1.333333
11	1.666667	2
13	2	2.333333
17	2.666667	3
19	3	3.333333
23	3.666667	4
25	4	4.333333
29	4.666667	5
31	5	5.333333
35	5.666667	6
37	6	6.333333
41	6.666667	7
43	7	7.333333

Table 38. Choosing an initial "a" value.

11 CONCLUSION 2—THE SIEVE FORMAT

It is observed that

    a. starting from gap base 2 to gap base 4, the difference is four for $f(x)$, for example observe $a_0 = 0$ and $a_0 = 1$ values;
    b. starting from gap base 4 to gap base 2, the difference is two for $f(x)$, for example $a_0 = 4$ for $f(x) = 23$ and $f(x) = 25$.

The difference will always be either four or two. Therefore, we can choose the type of sieve structure by making a choice on the values of a_0. If we choose, for example, $a_0 = 1$ for $f(x) = 5$ and $a_0 = 1$ for $f(x) = 7$, then our prime number-generation function becomes

    a. $f(x) = 6x + 5$ for gap base 4
    b. $f(x) = 6x + 7$ for gap base 2.

This produces a sieve where for a given value of x, then the difference is two. As a result, this type of sieve is called the twin-prime gap classification sieve. This would require a different approach in terms of equations for primality as well as sieving. This type of classification starts with gap base 4 and hence it is not the preferred standard for generation equations. Now in a similar manner, $6x \pm 1$ would give us a twin-prime structure though it would keep the gap base 2 and 4 sequence.

If all the elements of gap base 2 and 4 were arranged in sequence, then we observe an interesting 2-4 consistent pattern. It is this type of pattern that makes pseudoprimes to be very special rather than being a side issue in respect to the prime numbers.

f(x)	Difference
1	2
5	4
7	2
11	4
13	2
17	4
19	2
23	4
25	2
29	4

THE THEORY OF PRIME NUMBER CLASSIFICATION

31	2
35	4
37	2
41	4
43	2
to infinity . . .	

Table 39. The difference relation.

This also explains why the sieve will either be a two-difference or a four-difference in terms of its structure between gap base 2 and 4. For consecutive numbers in a gap base, this still gives us the difference of 2 + 4 = 6. It actually illustrates one of the arguments raised that the prime numbers are best dealt with in the context of underlying sets, which tend to show greater relationship than the universal set.

12 The Twin-Prime Conjecture

The twin-prime conjecture states that there are an infinite number of twin primes $(p, p+2)$, but it has been a challenge to prove the case. An approach can be made by using the sieve properties. Since there are two different types of sieve structures, any one of them can be used to define the framework for the proof.

Theorem

Let p_a be a prime number, then from p_a there exists

a. a previous prime p_2 such that $p_2 = p_1 + 2$
b. a next prime p_4 such that $p_4 = p_3 + 2$
where p_1 and p_3 are prime.

This has been stated slightly different from the conventional sense, which states that for a given largest twin prime, there exists a next twin-prime pair. The notion here is that there is always a number p_a that has a previous prime pair—that is definite. The end result is the same since both seek to establish that indeed there is a next twin-prime pair.

To construct the framework for the proof, an example can be made from $p = 13$. Then there is a previous prime pair $(5, 7)$ and a next prime pair $(17, 19)$. Hence, $p_2 = 7$ and $p_4 = 19$. This is the basic notion for the proof.

THE THEORY OF PRIME NUMBER CLASSIFICATION

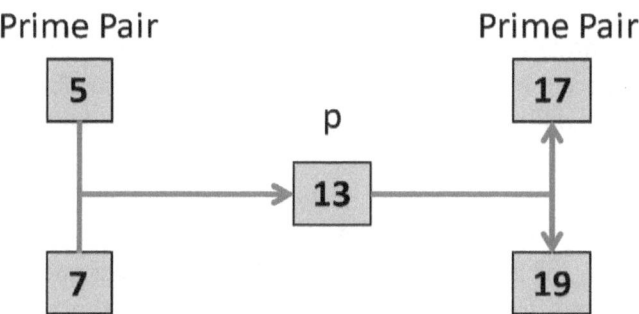

Figure 77. Framework for constructing twin-prime proof.

The proof also makes use of the fact that it has already been proved that there are an infinite number of primes; therefore, the prime p can be chosen to exist at infinity. Secondly, the argument is not whether twin primes exist at infinity or not but whether indeed we can guarantee the existence of the next twin prime at infinity. This is a credible argument given that the number of primes gets smaller and smaller as one gets to infinity. The key part of the proof lies in the consistency of the pattern for a twin-prime occurrence in the sieve.

Proof

Consider the occurrence of a twin prime in the sieve. Taking values of x and $(x + 1)$, we can construct a table.

Value	Gap Base 2	Gap Base 4
x	$f(x)$	$f(x) + 4$
x+1	$f(x + 1)$	$f(x + 1) + 4$

Now $f(x + 1) - f(x) + 4$ is the gap we seek to establish. But

$$f(x + 1) = f(x) + 6$$

Hence, we get $f(x + 1) - f(x) + 4 = 6 - 4 = 2$. A twin prime then will always occur when $f(x) + 4$ and $f(x + 1)$ are prime numbers. Hence, the twin prime will always start at gap base 4. That is, p_1 is at gap base 4, then $p_1 + 2$ will always be in gap base 2.

12 THE TWIN-PRIME CONJECTURE

At random, let there be a prime number p_a in the x-sieve at $x = n_1$, where n_1 is at infinity and $p_a - 2$ is not prime. Then the previous prime number p_2 in gap base 2 will occur at $p_2 = p - 6h_1$, where h_1 is the number of primes and pseudoprimes and is comparatively small in the context of infinity. Therefore, for p_2, then $x_1 = n_1 - h_1$.

Similarly, let there be a prime number q in the y-sieve that is a shadow prime of p_a at $x = m_1$, where $m_1 < n_1 - h_1$ and $q - 2$ is a pseudoprime. Then let there be a previous prime number p_1 in gap base 4 such that $p_1 = m_1 - 6h_2$, where h_2 is the number of primes and pseudoprimes. Therefore, for p_1 then $x_2 = m_1 - h_2$. Even though p_a is located in a random manner, but q is relative to p_a because it is a shadow prime since $m_1 < n_1 - h_1$.

If p_2 is the next prime of p_1, then $(n_1 - h_1) - (m_1 - h_2) = 2$ since this is a property of the gap-base structure. Hence, $p_2 = p_1 + 2$, and $(p_1, p_1 + 2)$ is the previous prime pair of p_a. Since p_a is chosen at random at infinity, and h_1 is assumed relatively small, then the previous prime pair of p_a is also at infinity.

Assume that (p_1, p_2) is the last prime pair. But the sieve structure is defined such that

   a. gap base 2 and gap base 4 form a parallel structure and not a sequential structure
   b. and the sets of gap base 2 and gap base 4 independently go to infinity.

Therefore, let there be a prime number $p(y)$ at infinity in gap base 2, then there exists a next prime $p(y + 1)$ in gap base 2. Similarly, if there exists a prime $p(z)$ in gap base 4, then there exists a next prime $p(z + 1)$ in gap base 4. Now the two sets are connected by the value of x to determine the following pattern outcomes in regard to the occurrence of $p(y + 1)$ and $p(z + 1)$, provided $p(y + 1) - p(z + 1) = 2$. Each of these outcomes will result in a twin prime when they occur.

Outcomes	x-sieve	y-sieve
Outcome 1	Prime	Prime

248

THE THEORY OF PRIME NUMBER CLASSIFICATION

	Prime	Prime
Outcome 2	Pseudoprime	Prime
	Prime	Pseudoprime
Outcome 3	Pseudoprime	Prime
	Prime	Prime
Outcome 4	Prime	Prime
	Prime	Pseudoprime

That is, a prime pair occurrence is defined by the structure of the sieve; hence, (p_1, p_2) cannot be the last pair.

Consequently, let there be a prime number p_b such that $p_b > p_a$, and $p_b - 2$ is a pseudoprime. Therefore, p_b is also at infinity. We have shown that for a prime number at infinity, there exists a previous prime pair. Hence, the previous prime pair of p_b will be (p_3, p_4). Now let $p_4 > p_a$, then (p_3, p_4) is a prime pair after p_a. Now two possibilities exist:

a. That (p_1, p_2) is any prime pair before p_a, and (p_3, p_4) is any prime pair after p_a, then we can conclude that (p_3, p_4) is a twin prime after (p_1, p_2). We are therfore certain there exists a prime pair after (p_1, p_2).

b. That (p_1, p_2) is an immediate prime pair occurrence before p_a, and (p_3, p_4) is an immediate prime occurrence after p_a, then we can conclude that (p_3, p_4) is the next twin prime after (p_1, p_2).

Now consider another prime p_c where $p_c > p_b$. In a similar manner, then p_c has a previous prime pair (p_5, p_6). Let $p_5 > p_b$; hence, (p_5, p_6) is either a prime pair after (p_3, p_4), or it is the next prime of (p_3, p_4). The process is therefore repetitive. Consequently, we can continue this same process an infinite number of times where always $p_n > p_m$ at infinity to get the same result of a previous prime pair (p_i, p_j) of p_n and $p_i > p_m$.

Therefore, at infinity, after a given prime pair, there exists another prime pair and, precisely, a next twin-prime pair.

[End Proof] (11:22 a.m., 19 July 2010).

12 THE TWIN-PRIME CONJECTURE

What we also learn from the proof is that the thinness of prime numbers is due only to the gap increase, not to the occurrence of prime numbers. That is, the primes are infinite, but the decrease over a given space gives the impression of scarcity of prime numbers.

13 Epilogue

It may be unusual to having closing comments for mathematical text like it was some drama. However, this was a form of drama on my part in terms of the experience that I underwent as a result of engaging in this research activity. I would like to share as much as possible the life part of the story rather than just the outcome of the project for the simple reason that mathematics does have a human face and, secondly, for the fact that God is not separate from mathematics—He created the numbers, rational and irrational, as well as the property of being prime. My belief is that there cannot be a pattern in the universe without God, in as much as there cannot be technology, creativity, and mathematics without man. This is not an attempt to win anybody to become a believer, but because of what He has done for me on a personal level, I would like to give Him the glory that is due to Him.

The first thing that I would like to share is the fact that many times I made the conclusion that I have found the secret of the prime numbers. This claim I made to myself because all claims begin on a personal level and are later verified by others to become universal or standard. But many times I found that my claim is wrong or that the context of the claim is right, but the methodology is not strong enough to be conclusive. My process of validation was based mostly on testing for conceptual consistency. This I find to be a very powerful tool that one needs to develop in expository mathematics since we are dealing with new ideas, emerging meaning, and definitive interpretation.

Most of the time, what drove me to continue is that fact that the context of the claim made a lot of sense, and this made me look for better ways of developing meaning and interpretation in order to have better tools and methodology to express the idea. I also needed a lot of self-discipline not to dash to make my results known, but to be calm about any particular finding and to work it using different conditions to see if I could arrive at the same conclusion. Some of the comments that I got from rejections by journals also helped to improve the quality of my conclusions—refereeing is a necessary tool for checking validity of a thought. It is just that in practice, it appears as regressive and unhelpful unless you conform and become part of organized mathematics.

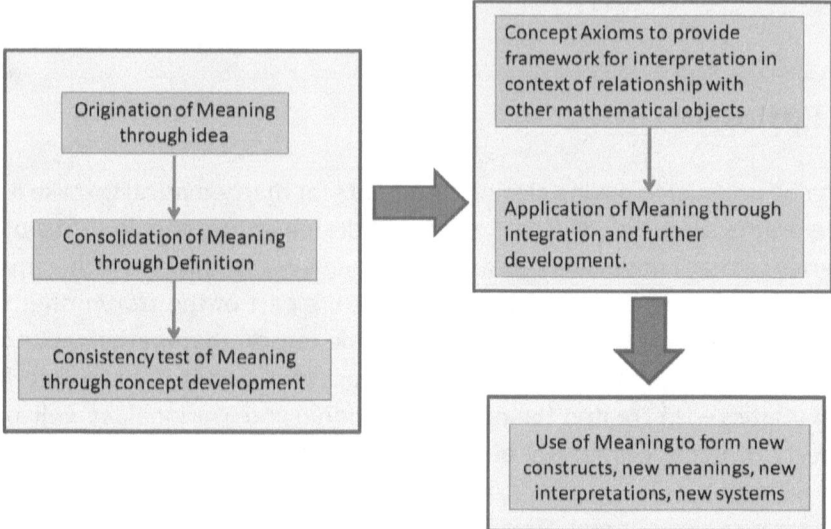

Figure 78. Schematic for meaning development and application in expository mathematics.

I also had to develop my mind, that is, in the following areas:

- The ability to think clearly and bring out ideas concisely.
- The ability to allow an idea to germinate and mature through research and further reading.
- I had to develop the science of creating constructs to express an idea and how constructs can be used to derive further ideas and knowledge.
- I had to learn how to discipline and make my imagination a resource for idea generation.

When I began this research, I did not ask myself the question, *what is a prime number?* I accepted the existing definition. I did not consider the meaning of the definition and its implications. It was only during the process that meaning became a consideration. My main concern was classification of prime numbers based on pattern methodology rather than algebraic expressions. I did look for meaning though, but it was meaning in patterns in order to develop classification techniques. However, this did not assist me in the development of how prime numbers are generated. This came accidentally, so to speak. I recognized certain properties of treating the prime numbers as a space and establishing a relationship between the prime number and its gap. It was at this moment that other questions began to arise, and something inside me was telling me that the way of prime number is conceptualized, is limited, or it can be revised. Imagination always works with something from inside called inspiration or even revelation—and these are not necessarily mathematical. This is where the plunge comes in, the point of divergence. If I was in a research program or in an institutional environment, it would have been very difficult to consider divergence of founding assumptions. Academic peer pressure would ensure that I remain consistent with what is already agreed on—the values and the traditions of thought and practice.

This makes me wonder then what types of conditions are conducive in order for expository mathematics to develop.

Notes

1. L. Garavaglia and M. Garavaglia, On the location and classification of all prime numbers.

2. R. Eismann, Decomposition of natural numbers into weight × level + jump and application to a new classification of prime numbers.

   http://arxiv.org/abs/0711.0865

3. R. K.Guy, *Unsolved problems in Number Theory*. New York. Springer-Verlag (2004).

4. S. Yates, *Periods of Unique Primes*, Mathematics Magazine, 53:5, p314, 1980.

5. C. K. Caldwell and Harvey Dubner, *Unique-period Primes*, J. Recreational Math., 29:1 (1998) 43-48.

6. Donald Gillies, Revolutions in Mathematics. Oxford Press. (1992)

Index

A

analysis
  horizontal, 100-101, 112, 121
  proportional, 124
  vertical, 121

B

between-gap prime count analysis, 111

C

chain reaction, 62
class, 44, 56, 134, 145
class distribution, 47
classification
  code, 186
  system, 22, 36, 50
  theory of, 9, 22, 162, 170
concept axioms, 191, 193
connecting arrow, 60
consecutive pseudoprimes, 23, 238
consequence, 58, 60, 163
count equation, 79
crossing group, 140, 142

Crowe
  seventh law of, 12
  tenth law of, 11

D

decomposition
  concept of, 34
  of natural numbers, 30
  positional, 34, 36
default prime space, 73
de Fermat, Pierre, 29
Delimitation, 16
delta analysis, 111-12, 115
Delta Classification System, 91
delta function, 94
delta prime
  gap, 92
  space, 67
delta variable space, 115
depth of the space, 69
Discrete Prime Space, 99, 101
distribution, 155
  pattern of, 53
  two facts of, 31
distribution analysis, 47
distribution theory of primes, 79

INDEX

ditation mathematics, 12

E

Eismann, 30, 255
Erdös, Paul, 30
even roots, 27, 37-38, 40-41, 62
events, 57-58
event signature, 58, 60-61
extrinsic classification, 43

F

Fermat primes, 29
fundamental gap theorem, 163, 183
  consequence of, 163, 165
fundamental theorem of arithmetic, 25, 27, 29

G

gap acceleration, 89
gap acceleration of the primes, 115, 123
gap analysis
  comparative, 112
  composite, 111-12, 115
  first-order singular, 111
  proportional, 112
  third-order singular, 112
gap base theory, 197
gap behavior, 21, 101, 132, 144, 153, 217
gap frequency graph, 108
gap in the gap, 89, 110, 115
gap theory
  of classification, 138
  of prime number classification, 133
  of prime number generation, 22

gap theory classification, 21, 129
generation, 8, 22, 26, 133-34, 148-49, 151, 153, 162-66, 169-72, 177, 179, 187, 201, 220-21, 239-42, 244
G-numbers, 178, 180
God, 14, 251
group behavior, 78

H

hypothesis
  1, 41
  2, 57

I

idea, 14-16, 18-19, 57-58, 62, 124, 188, 251-52
imagination, 16-18, 25, 252-53
impact, 58
infinity, 68
influence, 58, 60
interpretation, 15, 251
intrinsic classification, 43
Intuition, 17

L

law of primes
  consequence of, 238
  example of, 23, 218
left operation primes, 56
linear delta space, 165
linear relationship, 84, 133, 144, 159, 170, 226-27

M

mathematics

cumulative, 17-18
experiential, 17
expository, 8, 16-18, 173, 188-89, 251-53
meaning, 9, 15, 251-53
mean of primes, 83
Mersenne, Marin, 29
Mersenne primes, 29, 131
methodology, 9, 19, 22, 251, 253
minimum rule, 217
multivalued function, 169, 181, 187

N

natural gap, 92, 133
natural numbers, 25, 27, 132, 152, 190, 255
nondecomposing prime, 37-39, 51
normal function. *See* standard function

O

odd root, 32
original event, 58

P

P1-test, 22, 182, 210, 212
P2-test, 22, 210
pattern methodology, 233, 253
patterns, 8, 29, 53, 108, 199, 251, 253
  dual-pattern, 199
  prime patterns
    dual, 199
    single, 199
  twin-prime, 200
positional classification, 21, 67, 71-72
positional space, 67, 71

primality condition, 22, 212
primality test. *See* P1-test
prime acceleration, 115
prime class, 53, 55, 146
prime count, 81, 102
prime count function, 101
prime density, 145-47
prime dimension, 104
prime distribution, 26, 50, 89
prime family, 95, 109-10, 145, 155
prime gap, 92, 100-101, 218
  analysis techniques, 111
  axioms, 21, 132
prime generation, 159, 163-65, 187, 201
  axioms, 164
  condition, 164
  space, 165
prime number decomposition, 35, 38
prime number decomposition operation, 35
Prime Number Generation, 8, 22, 26, 148-49, 151, 153, 162-63, 169, 197
prime number-generation axioms, 22
prime number occurrence, 23, 36, 198
prime numbers
  axioms of, 190
  behavior of, 31, 144
  as building blocks, 25
  definition of, 29, 188-90
  existence of, 68
  as having prime signatures, 21
  law of, 23
  patterns of, 29, 215
  as prime roots, 35

and prime waves, 59
as pristine, 37
and pseudoroots, 39
qualification of, 62, 189-90
as random, 27, 61
as rare prime, 37, 41
as sequence, 65, 147
spectral lines of, 144-45
study of, 131
theorem of, 173
three types of, 140
prime pair, 12, 249
prime reaction, 63
prime root, 33, 35, 41-42, 44-46, 59, 62
prime root classification system, 14
prime root theory, 21
prime signature, 32
prime space
  concept of, 104
  default, 73
  definition of, 69, 92
  delta, 67, 93
  discrete, 99, 101
  span of, 92
  standard, 72
  structure of sequence of, 71
  universal, 76
prime space gap, 100-101
prime subspace, 68, 125
prime trace, 57
prime waves, 59
priority rule, 40-42
pristine primes, 37-38, 41
probabilistic event, 124
probability, 22, 110, 124-27, 186, 212, 221
probability gap analysis, 112
probability test. *See* P2-test

pseudohit, 201
pseudoprime, 164-65, 179, 215
pseudoroots, 39

R

random event analysis, 21
random function, 22, 157-60
randomization, 200
range, 47, 60
rare prime, 37
reaction, 62
Research Question, 25-27, 58, 65, 89, 129, 149, 175, 190
result, 58-60, 191
revolutions, 11, 13-14, 17, 255
right operation primes, 56
root sequence, 35
root unique prime, 37, 44

S

Selfridge, John, 30
sequencing, 31, 155, 184
sequential root decomposition, 36
shadow family, 98
shadow gap, 98
sieve
  algebraic sieve, 22, 187, 198, 204
    external properties of, 201
    internal dynamics of, 194
  axioms, 181, 190
  of Eratosthenes, 151, 163
  structure of the, 22
  theory of, 22, 177
  twin-prime gap classification, 244
signature, 57
singular analysis
  first order, 111

second order, 112
  third order, 112
singular gap, 110, 117, 133, 138
singular gap analysis. *See* gap in the gap
singular gap of the delta space, 136
span, 92
standard function, 169
standard prime, 37, 52
Standard Prime Space, 72
structure, 9, 31, 40, 43, 53, 71, 198

T

theorem, 9, 16, 125, 173, 238, 246
  1, 36, 40
  2, 44

consequence of, 38
theory of delta classification, 21, 71
theory of prime number decomposition, 27
twin-prime conjecture, 8, 23, 100, 246
twin-prime gap classification, 244
twin primes, 184, 197

W

Wave Influence, 60
Wave Interruption, 59

Z

Zagier, Don, 31
zero root, 38-39

www.ingramcontent.com/pod-product-compliance
Lightning Source LLC
Chambersburg PA
CBHW031832170526
45157CB00001B/272